Joint Source–Channel Video Transmission

Joint Source–Channel Video Transmission

Fan Zhai and Aggelos Katsaggelos

www.morganclaypool.com

ISBN: 1598290444   paperback
ISBN: 9781598290448   paperback

ISBN: 1598290452   ebook
ISBN: 9781598290455   ebook

DOI: 10.2200/S00061ED1V01Y200707IVM010

A Publication in the Morgan & Claypool Publishers series
SYNTHESIS LECTURES ON IMAGE, VIDEO, AND MULTIMEDIA PROCESSING #10

Lecture #10
Series Editor: Alan C. Bovik, University of Texas at Austin

Library of Congress Cataloging-in-Publication Data

Series ISSN: 1559–8136   print
Series ISSN: 1559–8144   electronic

First Edition
10 9 8 7 6 5 4 3 2 1

Joint Source–Channel Video Transmission

Fan Zhai
Texas Instruments

Aggelos Katsaggelos
Northwestern University

SYNTHESIS LECTURES ON IMAGE, VIDEO, AND MULTIMEDIA PROCESSING #10

MORGAN & CLAYPOOL PUBLISHERS

ABSTRACT

This book deals with the problem of *joint source-channel video transmission*, i.e., the joint optimal allocation of resources at the application layer and the other network layers, such as data rate adaptation, channel coding, power adaptation in wireless networks, quality of service (QoS) support from the network, and packet scheduling, for efficient video transmission. Real-time video communication applications, such as videoconferencing, video telephony, and on-demand video streaming, have gained increased popularity. However, a key problem in video transmission over the existing Internet and wireless networks is the incompatibility between the nature of the network conditions and the QoS requirements (in terms, for example, of bandwidth, delay, and packet loss) of real-time video applications. To deal with this incompatibility, a natural approach is to adapt the end-system to the network. The joint source-channel coding approach aims to efficiently perform content-aware cross-layer resource allocation, thus increasing the communication efficiency of multiple network layers. Our purpose in this book is to review the basic elements of the state-of-the-art approaches toward joint source-channel video transmission for wired and wireless systems.

In this book, we present a general resource-distortion optimization framework, which is used throughout the book to guide our discussions on various techniques of joint source-channel video transmission. In this framework, network resources from multiple layers are assigned to each video packet according to its level of importance. It provides not only an optimization benchmark against which the performance of other sub-optimal systems can be evaluated, but also a useful tool for assessing the effectiveness of different error control components in practical system design. This book is therefore written to be accessible to researchers, expert industrial R&D engineers, and university students who are interested in the cutting edge technologies in joint source-channel video transmission.

KEYWORDS

Video processing, video transmission, joint source-channel coding, resource allocation, error resilience, unequal error protection (UEP), error control, quality of service (QoS), multimedia streaming, cross-layer optimization, content-aware video transmission.

Contents

List of Figures

List of Tables

CHAPTER 1

Introduction

The essence of telecommunications is to exchange information over long distances, by such means as mail, telephone, radio, satellite, television, and the Internet. On May 24, 1844, Samuel Morse sent his first public message over a telegraph line between Washington and Baltimore, which opened a new page in the history of modern telecommunications.

With the advancement of technologies in the areas of computers, wireless communications, semiconductors, and material science, telecommunications, among others, are continuously changing face and objectives. New applications appear everyday. People can now enjoy the exchange of a variety of signals, such as text, audio, pictures, and video. In addition, information exchange has been significantly faster and considerably less costly than ever before. Thus, the trend in modern telecommunications is to allow users to exchange new types of content and to significantly increase the amount and speed of information exchange. Thanks to the development of the Internet and wireless communication infrastructure, modern telecommunications allow information exchange among "any one", at "any time", and "any where".

This monograph addresses the topic of video communications. Compared to other types of signals, such as text, audio, and images, a key characteristic of video signals is the vast amount of data they require for their representation. For example, the bandwidth required by the National Television System Committee (NTSC) for network quality video is about 45 megabits per second (Mbps) [7]. Recommendation 601 of the International Radio Consultative Committee (CCIR) calls for a 216 Mbps bandwidth for video objects. A video object based on the HDTV (High Definition Television) quality images requires at least 880 Mbps bandwidth. Such an amount of data places a high demand on the transmission and storage requirements. Another key difference between video communications and more traditional data communications is that video transmission applications usually impose a strict end-to-end delay constraint. Due to the technological advancements that enable wider bandwidth, higher throughput, and more powerful computational engines, video transmission applications, such as video broadcasting, videoconferencing, distance learning, and on-demand video streaming, have gained increased popularity. In addition, the demand for faster and location-independent access to multimedia sources offered by the current Internet is steadily increasing. With the deployment of higher data rate cellular networks, such as GSM-GPRS, UMTS, and CDMA 2000, and the proliferation of cameras and video-display-capable mobile devices, providing video capabilities to consumers over the cellular communication infrastructure can generate a large number of valuable applications.

1.1   APPLICATIONS

It is important to classify video transmission applications into groups, since the nature of a video application determines the constraints and the protocol environment the video source coding has to cope with. Based on the delay constraints, video transmission applications can be generally classified into the following three categories [8].

- *Conversational applications*: These applications include two-way video transmission over ISDN, Ethernet, LAN, DSL, wireless and mobile networks, such as videotelephony, distance learning, and videoconferencing. Such applications are characterized by very strict end-to-end delay constraints, usually less than a few hundred milliseconds (with 100 ms as the goal). They are limited to point-to-point or small multipoint transmissions. They also implicitly require the use of real-time video encoders and decoders, which allows the tuning of the coding parameters in real time to appropriately adjust the error resiliency of the bitstream, and often the use of feedback-based source coding tools. Source coding adaptation usually closely interacts with channel coding adaptation for this type of applications. On the other hand, the strict delay requirements usually limit the allowable computational complexity, especially for the encoder. Strict delay constraints further prevent the use of some coding tools that are optimized for high-latency applications, such as bipredicted slices.

- *Video download and storage applications*: For video download applications, the video source is usually preencoded and stored on a server. The application treats the encoded video bitstream as a regular data file, and uses reliable protocols such as FTP or HTTP for the downloading. Since the video encoding process does not have to be in real time, computational complexity of the encoder is less of an issue. Hence, the video coding can be optimized for the highest possible coding efficiency, and does not have to consider constraints in terms of delay and error resilience. Video storage can also be placed into this category, since it also has relaxed computational complexity and delay constraints, and similarly, error resiliency is not a concern. This type of application is the focus of most of the traditional video coding researches, where improving the compression efficiency is the ultimate goal.

- *Video streaming applications*[1]: This type of video application is somewhere in between download and conversational applications, with respect to the delay characteristics. Unlike video download/storage application, these applications allow the start of playback before the whole video bit stream has been transmitted. The initial setup/buffering time is usually a few seconds. Once the playback starts, it must be real time, which means that it must be continuous and without interruption. The video stream can be either preencoded and transmitted on demand, or a live session compressed on the fly. The video to be streamed is sent from a single server, but may be distributed in a point-to-point, multipoint, or even broadcast fashion.

  Under normal conditions, streaming services use unreliable transmission protocols, which we will discuss in detail in Chapter 2. Thus, as with the conversational applications, error-resilient source coding is generally required. Compared to conversational applications, however, very limited interaction of source and channel coding can be employed in streaming applications. This is because for this type of applications, e.g., broadcast applications, the encoder usually has only limited, if any, knowledge of the network conditions. Because different users may have different connection quality to the server, the encoder has to adapt the error resilience tools to a level acceptable by most users. In addition, due to the relaxed delay constraints in this case, compared to conversational services, certain high-delay video coding tools, such as bipredicted slices, may be applicable.

In addition to the above-mentioned categories, developing wireless technologies, such as third-generation (3G) and the emerging fourth-generation (4G) cellular systems, IEEE 802.11 wireless local area network (WLAN) standards, and Bluetooth systems, are expanding the scope of existing video applications and are leading to the creation of new ones. One such example is multimedia messaging services (MMS) over ISDN, DSL, Ethernet, LAN, wireless and mobile networks. Moreover, new applications may be deployed over existing and future networks.

This monograph deals primarily with conversational video applications, as they are likely to be the most challenging ones. As discussed above, this type of application requires a close interaction between source and channel coding, which is the focus of this monograph. We also discuss certain streaming applications, in particular, point-to-point live video streaming with relatively short end-to-end delay, e.g., video streaming over LAN. This type of application has

---

[1]Note that there is no generally accepted definition for the term "streaming"; here we adopt the most commonly used one [8].

the same level of delay constraint as conversational applications, thus calling for the applications of the same source and channel coding techniques.

Video communications is a very broad topic. From the system design point of view, it consists of an end system at the sender side, a communication channel, and an end system at the receiver side. In this monograph, we focus on the end system design at the sender side. The system design usually consists of three major elements: video coding and decoding (codec), rate control or congestion control, and error control. The video codec consists of an encoder located at the sender side and a decoder located at the receiver side. Rate control is used to judiciously allocate bits/bandwidth between and within video frames to maximize the overall communication quality. Error control refers to the techniques used to combat channel errors. In this monograph, we will briefly discuss video codecs and rate control, but the focus is on error control. Next, we discuss the technical challenges of video transmission, the general approach toward error control, and the state-of-the-art techniques in implementing high quality error control to achieve the best video transmission quality by efficiently utilizing the network resources.

1.2   TECHNICAL CHALLENGES

Video communications face various technological challenges. The first one is the quality of service (QoS) mismatch between the video applications and the communications networks. For video applications, the QoS at the application layer is a function of the quality of the delivered video, which can be measured either quantitatively in terms of a distortion metric or subjectively. The quality of the delivered video is determined by the QoS provided by the network, which is usually defined in terms of bandwidth, probability of packet loss, and delay distribution.

In a video communications system, the video is first compressed and then divided into packets with fixed or variable length and multiplexed with other types of data, such as audio. To achieve acceptable delivery quality, the transmission of a video stream typically requires a relatively high bandwidth. Unless a dedicated link that can provide a guaranteed QoS is available between the source and the destination, data bits or packets may be lost or corrupted, due to either traffic congestion or bit errors caused by the impairments of the physical channels. Such is the case, for example, with the current Internet and wireless networks. Owing to its best effort design, the current Internet makes it difficult to provide a guaranteed QoS in terms of bandwidth, packet loss probability, and delay imposed by the video communication application. Compared to wired links, wireless channels are much noisier because of fading, multipath, and shadowing effects, which results in a much higher bit error rate (BER) and consequently an even lower throughput [9,10]. In wireless networks, according to the current standards, a packet with unrecoverable bit errors is usually discarded at the link layer.

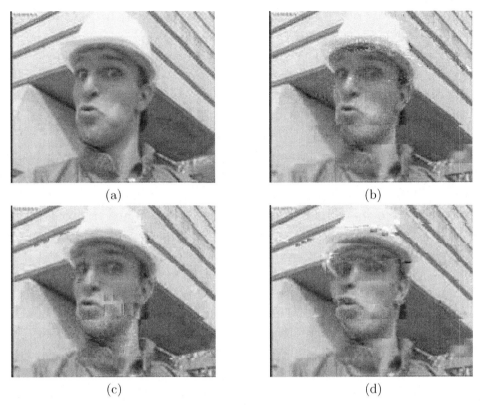

FIGURE 1.1: Illustration of the effect of channel errors on a video stream compressed using the H.263 standard (a) original frame; reconstructed frame with (b) 3% packet loss, (c) 5% packet loss, (d) 10% packet loss, (QCIF foreman sequence, frame 90, coded at 96 kbps and frame rate 15 fps, each packet is one row of macroblocks).

Compressed video streams are sensitive to transmission errors because of the use of predictive coding and variable-length coding (VLC) by the source encoder. Due to the use of spatio-temporal prediction, a single-bit error can propagate spatially and temporally. Similarly, because of the use of VLCs, a single-bit error can cause the decoder to loose synchronization, so that even successfully received subsequent bits may become unusable. Figure 1.1 illustrates the effect of channel errors on a typical compressed video sequence in the presence of packet loss. In the simulation, QCIF (176×144) Foreman test sequence was used, and each row of macroblocks was encoded as one packet. Error concealment techniques, which will be discussed in detail in Chapter 2, were utilized at the decoder to mitigate the effect of packet loss.

In addition, packets that arrive at the receiver after a defined deadline will usually be considered as lost, since they may not be useful for playback. This excessive delay is usually caused by network congestion, which results in longer queueing delays and even overflow at the buffers of the routers and switches.

In a network without QoS guarantee, packet loss and delay are both random processes. The implication of this is not as severe for traditional applications such as data transfer and electronic mail. This is because every type of application has its own type of quality impairment under the same network conditions, and therefore calls for different QoS. For example, "elastic" applications, such as web browsing, data file transfer, and electronic mail, are not sensitive to delay. However, for the deployment of delay-sensitive (but packet-loss tolerant) multimedia applications, the lack of QoS guarantees, as in the current Internet and wireless networks, has major implications [11, 12]. This is because, for elastic applications, reliable transmission can always be achieved through retransmission, while for real-time video applications, retransmission-based techniques may not be applicable due to the tight delay constraints.

Reliable real-time video transmission usually requires error control which aims at limiting the effects of channel errors and therefore improving the received video quality. However, as both the information rate provided by the video source and the channel conditions are time varying, it is not possible to derive an optimal solution for a specific transmission of a given video signal. Furthermore, the network resources are limited and may vary with both time and location. Examples of network resources, a general term in communication networks, are transmission bandwidth, buffers at the routers and switches, buffers at the sender or the receiver end, the computational capability of the encoder, decoder, and transcoder, transmission cost in networks with pricing capabilities, and transmission power and battery life in wireless communications. Some of the constraints on resources, such as the buffer size, computational speed, and display resolution at the user end, are *hard*, while other constraints are *soft* in the sense that they aim at making the system efficient or treating other users in the communication system fairly. One example of a soft constraint is the TCP-friendly protocol used to perform congestion control for media delivery applications, where the source bit rate is constrained so that various types of traffic can fairly share the network resources [11]. Certain constraints, such as the computational capability of the system, are extremely important limiting factors for video applications, especially for conversational applications, as a great amount of computations has to be performed in a short time period due to the strict delay constraint.

Although technologies relaxing such constraints, for example, technologies providing higher bandwidth, more powerful computational capabilities, and longer battery life, are continuously being deployed, better QoS will inevitably lead to higher user demands of service. Thus, the topic of this monograph is how to design an end system that is capable of optimally utilizing the limited resources to achieve the best video delivery quality.

As mentioned above, to address these technical challenges, we need to enforce error control. Error-control techniques, in general, include error-resilient source coding, forward error correction (FEC), retransmission, power control, network QoS support, and error conceal-ment. To maximize the error control efficiency, limited network resources should be optimally allocated to video packets, which typically requires joint consideration of source and channel coding.

1.3 JOINT SOURCE-CHANNEL CODING/(CROSS-LAYER RESOURCE ALLOCATION)

Due to the "unfriendliness" of the channel to the incoming video packets, these packets have to be protected so that the best possible quality of the received video can be achieved at the receiver's end. A number of techniques, which are collectively referred to as error resilient techniques, have been devised to combat transmission errors. They can be grouped into [13]: (i) those introduced at the source and channel coder to make the bitstream more resilient to potential errors; (ii) those invoked at the decoder upon detection of errors to conceal the effects of errors, and (iii) those which require interactions between the source encoder and decoder so that the encoder can adapt its operations based on the loss conditions detected at the decoder.

Traditional source coding aims at removing the redundancy of the source in order to achieve the rate-distortion limit of the source or the operational rate-distortion limit of the codec [14–17]. While such an objective results in codecs with significant source-coding performance, it does not adequately address video communications over hostile channels. The development of error resilient approaches or approaches for increasing the robustness of the multimedia data to transmission errors is a topic of utmost importance and interest.

To make the compressed bitstream resilient to transmission errors, redundancy must be added into the stream. Such redundancy can be added either by the source or the chan-nel coder. Conventional video communication systems have focused on video compression, namely, rate-distortion optimized source coding [16]. Shannon said over fifty years ago [18,19] that source coding and channel coding can be separated for the optimal performance of a communication system. The source coder should compress a source to a rate below the chan-nel capacity while achieving the smallest possible distortion, and the channel coder can add redundancy (e.g., through FEC) to the compressed bitstream to enable the correction of trans-mission errors. Following Shannon's separation theory, there have been major advances in source coding (e.g., rate-distortion optimal coders and advanced entropy coding algorithms) and channel coding (e.g., Reed–Solomon codes, Turbo codes, and Tornado codes). The sep-aration theory promises that the separate design of source and channel coding not only does not introduce any performance sacrifices, but can also greatly reduce the complexities of a prac-tical system design. However, the assumptions on which separation theory is based (infinite length codes, delay, and complexity) may not hold in a practical system. This leads to the development of the approach of joint consideration of source and channel coding, referred to as *joint source-channel coding* (JSCC). JSCC can greatly improve the system performance when there are, for example, stringent end-to-end delay constraints or implementation complexity concerns.

In addition, one of the main characteristics of video is that different portions of the bitstream have different importance in their contribution to the quality of the reconstructed video. In other words, not all bits are equal in a compressed video stream. For example, in an MPEG video bitstream, I (intra) frames are more important than P (predictive) and B (bidirectional predictive) frames. If the bitstream is partitioned into packets, intracoded packets are usually more important than intercoded packets. If error concealment is used, the packets that are hard to conceal are usually more important than easily concealable ones. In a scalable video bitstream, the base layer is more important than the enhancement layer. Therefore, unequal error protection (UEP) is a natural choice in video transmission. UEP can enable a prioritized protection of video packets through some error-resilient source coding tools, such as data partitioning and slice structure. UEP can also be realized through prioritized transmission by techniques based on different levels of FEC and/or retransmission, transmitter power adaptation, or DiffServ (Differentiated Services) [1,20,21]. Joint consideration of source coding and channel coding can maximize the efficiency of UEP.

Therefore, recent video coding research has focused on the joint design of end-system source coding with manipulations at the other layers, such as channel coding, power adaptation in wireless networks, and QoS support from the network [e.g., differentiated services networks and integrated services networks (IntServ)], to name a few [22]. When extending the discussion from a simple wired or wireless channel to a network, a more general term for JSCC can be used, i.e. "joint source and network coding" (JSNC), which may include packet scheduling, power adaptation, etc. In this monograph, we use the term "channel encoding" in a general sense to include modulation and demodulation, power adaptation, packet scheduling, and data rate adaptation, and henceforth we will utilize the term JSCC in this context to encompass JSNC.

1.4 SCOPE AND CONTRIBUTIONS

In order to provide the readers with the big picture as to where the topic of this monograph belongs to, we start from the classical open systems interconnection (OSI) seven-layer communication model. Figure 1.2 illustrates such a model, where layer 5 (session layer), layer 6 (presentation layer) and layer 7 (application layer) are merged into one application layer to better describe our work. The resulting five-layer model is also referred to as a TCP (transport control protocol) model. The general functions of these layers are as follows.

Layer 5, 6, 7: The application layer—This layer provides applications services to user and programs, such as file transfers, http, electronic mail, etc.

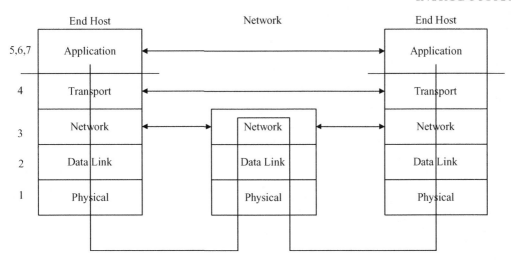

FIGURE 1.2: OSI networking model.

Layer 4: The transport layer—This layer provides transparent transfer of data between hosts and is responsible for end-to-end error-correction and flow control. TCP and UDP (user data protocol) work at this level.

Layer 3: The network layer—This layer deals with network addressing and routing of data between two hosts and any congestion that might develop.

Layer 2: The link (or data link) layer—This layer defines the format of data on the network. In IP-based wireless networks, the link layer is divided into two sublayers: the MAC (medium access control) layer and LLC (logical link control) layer. The MAC sublayer is responsible for communications between stations by coordinating medium access. The LLC sublayer manages frame synchronization, error correction, and flow control.

Layer 1: The physical layer—It deals with physical aspects such as physical media, electrical impulse, transmitter power, modulation and etc.

As the video encoder and decoder are located at the application layer, our focus in this monograph is on the interaction between the video encoder and the underlying layers. Thus the decoder-based techniques, such as post-processing, joint source-channel decoding [23], and receiver-driven channel coding [24, 25] are not discussed here. They are complementary to the objective of this monograph.

The traditional layered protocol stack, where various protocol layers can only communicate with each other in a restricted manner, has proved to be inefficient and inflexible in adapting to the constantly changing network conditions [26]. For the best end-to-end performance, multiple protocol layers should be jointly designed and should be able to react to the channel conditions in order to make the end system *network adaptive*. Such an approach, referred to collectively as *cross-layer* design [26–29], represents a very active topic of research and development.

With respect to cross-layer design at the sender side, traditionally, there are two approaches to address the challenges described above for video communications. In the networking community, the approach is to develop protocols and mechanisms to adapt the network to the video applications. One example of such an approach is to modify the mechanisms implemented in the routers/switches to provide QoS support to guarantee the required bandwidth, bounded delay, delay jitter, and packet loss (such as DiffServ and IntServ) for video applications. Another example is to use additional components such as an overlay network, an edge proxy, which provides application layer functions such as transcoding, rate adaptation, and FEC. Approaches toward cross-layer design followed by the networking community, such as the ones mentioned above, are not addressed in this monograph.

In the video processing community, on the other hand, the approach toward cross-layer design is to adapt the end system to the network, which is what we employ in this monograph. We assume that the lower layers provide a set of prespecified adaptation components; from the encoder's point of view, these components can be regarded as network resource allocation "knobs". Based on the assumption that the encoder can access and adjust those resource allocation knobs, we present a general resource-distortion optimization framework, which assigns network resources from multiple layers to each video packet according to its level of importance.

Depending on the available adaptation components, this framework is manifested in different forms. For example, for Internet video transmission, we can jointly design source and channel coding (FEC and/or retransmission), while for video transmission over wireless networks, we can jointly consider source–channel coding, power adaptation, and scheduling. The framework provides not only a benchmark against which the performance of other suboptimal systems can be evaluated, but also a useful tool for assessing the effectiveness of different error control components in practical system design. It is used throughout the monograph to guide our discussions when reviewing the state-of-the-art approaches toward JSCC for wired and wireless systems alike, for the current, recently proposed, and emerging network protocols and architectures.

The rest of the monograph is organized as follows. In Chapter 2, we first provide an overview of a video transmission system. We then present the basic idea of JSCC, starting with basic rate-distortion definitions in Chapter 3. Following that, we describe the basic components of error-resilient video coding in Chapter 4, and channel coding techniques that are widely used for video communications in Chapter 5. In Chapters 6 and 7, we address video communications over the Internet and wireless networks, respectively. Finally, Chapter 8 contains concluding remarks.

CHAPTER 2

Elements of a Video Communication System

This chapter provides the general background of this monograph. It is intended to provide to the readers the basic information on how compressed video bitstreams are packetized and transmitted through multiple network layers. We first provide a brief high-level overview of a video transmission system, followed by a discussion of the details of network interfaces. As error control is the focus of this monograph, we then give a brief overview of the major error control tools that sit in different places of a video transmission system, from the source encoder, channel coder, transmission network, to the source decoder. At the end, we discuss the video quality metrics used to evaluate the performance of a video communication system. The two components this monograph is primarily focused on, source encoder and channel encoder, are discussed in detail in Chapters 4 and 5, respectively.

2.1 VIDEO COMMUNICATION SYSTEM

The block diagram of a video communication system is shown in Fig. 2.1, which represents a simplified version of the OSI reference model shown in Fig. 1.2. Compared to the general OSI model, the key components of a video communication system, including video encoder, video decoder, and rate control, are illustrated in Fig. 2.1. Figure 2.1, however, does not include the lower three layers, which we will cover later in this chapter. As our goal here is to provide a high-level overview of video communication systems, such a simplified model is adequate for our discussion at this point.

The system in Fig. 2.1 has the following five major conceptual components: (1) the source encoder that compresses the video signal (the multimedia source in general) into media packets, which are sent directly to the lower layers or are uploaded to the media server for storage and later transmission on demand; (2) the application layer in charge of channel coding and packetization; (3) the transport layer that performs congestion control and delivers media packets from the sender to the receiver for the best possible user experience, while sharing network resources fairly with other users; (4) the transport network which delivers packets to

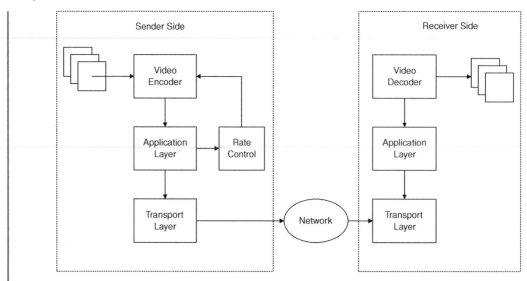

FIGURE 2.1: Video transmission system architecture.

the client; and (5) the receiver that decompresses and renders the video packets and implements the interactive user controls based on the specific applications [1, 22].

The compression or coding of a signal (e.g., speech, text, image, video) has been a topic of great interest for a number of years. Numerous results and successful compression standards exist (e.g., MPEG-2, MPEG-4, and H.264). Source compression is the enabling technology behind the multimedia revolution we are experiencing. The objective of compression is to reduce the source redundancy. Lossless compression aims at compressing the source at a rate close to the entropy of the source. Since the bit rate achieved using lossless compression is usually much higher than the available channel capacity, lossy compression is generally required for video transmission applications. Most practical communication networks have limited bandwidth and are lossy by nature. Facing these challenges, the *video encoder* aims at efficiently compressing the original video sequence in a lossy fashion while maintaining the resiliency of the bitstream to channel errors. These two requirements are conflicting and they establish the tradeoff between source and channel encoding. Compression reduces the number of bits used to represent the video sequence by exploiting both temporal and spatial redundancy (as well as properties of the human visual system when applicable). On the other hand, to mitigate the effect of channel errors on the decoded video quality, redundancy is added back to the compressed bitstream during channel encoding. This topic is detailed in Chapter 4.

The source bit rate is shaped or constrained by a rate controller that is responsible for allocating bits to each video frame and each video unit (such as a macroblock) within a frame or to each video packet. This bit rate constraint is set based on the channel rate and the estimated channel state information (CSI) reported by the lower layers, such as the application and transport layers. As already mentioned, we will look at the network, data link, and physical layers, sitting below the transport layer later in this chapter, since allowing the various layers to exchange information leads to a cross-layer design of a video communication system, which is a central theme in this monograph.

In Fig. 2.1, the *network* block represents the communication path between the sender and the receiver. This path may include routers, subnets, and wireless links. The network may have multiple channels (e.g., a wireless network) or paths (e.g., a network with path diversity), or support QoS (e.g., integrated services or differentiated services networks). Packets may be dropped in the network due to congestion, or at the receiver due to excessive delay or unrecoverable bit errors in a wireless network. To combat packet losses, parity check packets, used for FEC, may be generated at the application/transport layer. In addition, lost packets may be retransmitted if the application allows.

For many source-channel coding applications, the exact details of the network infrastructure may not be available to the sender and they may not always be necessary. Instead, what is important in JSCC is that the sender has access to or can estimate certain network characteristics, such as the probability of packet loss, the transmission rate, and the round-trip-time (RTT). In most communication systems, some form of CSI is available to the sender, such as an estimate of the fading level in a wireless channel or the congestion over a route in the Internet. Such information may be fed back from the receiver and can be used to aid in the efficient allocation of resources.

At the receiver side, the transport and application layers are responsible for de-packetizing the received transport packets, channel decoding (if FEC is used), and forwarding the intact and recovered video packets to the video decoder. The video decoder then decompresses the video packets and displays the resulting video frames in real-time (i.e., the video is displayed continuously without interruption at the decoder). The video decoder typically employs error concealment techniques to mitigate the effect of packet loss. The commonality among all error concealment strategies is that they exploit spatio-temporal correlations in the received video sequence to conceal lost information.

2.2   NETWORK INTERFACE

The module between applications and networks, known as the network interface, consists of all five major layers, namely, the application layer, the transport layer, the network layer, the link layer, and the physical layer. The main functionality of the network interface is to packetize the compressed video bitstream and then send these packets to the receiver through the network by a certain deadline imposed by the application at a constrained rate based on the estimation of the network conditions. Common issues concerning the network interface include packetization, channel coding including retransmission, congestion control, and network condition monitoring [22].

The network interface serves as the foundation and core of joint source-channel coding. Specifically, the performance of a JSCC scheme heavily relies on the accuracy of the estimation of the time-varying channel conditions and the full interaction of different layers. The latter enables efficient forwarding of lower layer QoS parameters to the upper layers and use of the application-layer QoS parameters (such as the level of each packet's contribution to the overall video delivery quality) to guide the transmission priority (prioritized transmission is used in schemes such as FEC, retransmission, and power adaptation, to name a few). Recent research on video transmission system design has been focusing on this network interface part [6,30–33].

In this section, we first briefly introduce the network protocols employed at multiple layers, and then focus on the major functions provided by the network interface in supporting real-time video transmission.

2.2.1   Network protocols

Currently, the most commonly used network layer protocol is the Internet Protocol (IP). It provides connectionless delivery services, which means that each packet is routed separately and independently regardless of its source or destination [34]. In addition, IP provides best effort and thus unreliable delivery services. Figure 2.2 shows the protocol stack used for an IP network. Communication protocols represent a rather wide topic which is beyond the scope of this monograph. As the objective of this monograph is video transmission, the application-layer

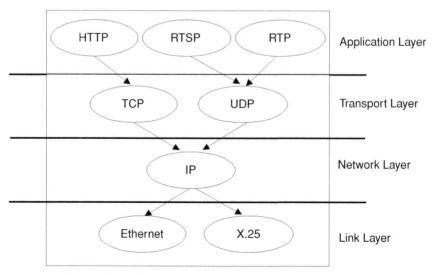

FIGURE 2.2: Illustration of protocol layers.

protocol, HTTP, and link-layer protocols, Ethernet and X.25, are not addressed herein. We focus instead on the part that is closely related to real-time applications.

The transport control protocol (TCP), operating at the transport layer, is a connection-oriented and reliable service [35]. TCP provides reliability by means of a window-based positive acknowledgement (ACK) with a go-back-N retransmission scheme. Note that TCP is one of the few transport-layer protocols that has its own congestion control mechanisms.

An alternative to TCP at the transport layer is the user datagram protocol (UDP), which together with IP is sometimes referred to as UDP/IP. Like TCP, UDP is connection-oriented. Unlike TCP, however, UDP does not provide reliable transmission. First, it does not provide sequencing of the packets as the data arrive. This means that the application that uses UDP must be able to ensure that the entire message has arrived and is in the right order. Second, UDP does not enforce retransmission of the lost packets. TCP, however, introduces unbounded delay due to persistent retransmissions, which may not be suitable for applications that have strict delay constraint. For this reason, UDP is widely used in real-time video applications.

UDP also differs from TCP in that it does not have its own congestion control. Thus, when UDP/IP is used, additional congestion control needs to be deployed on top of UDP to constraint the bit rate for the applications. This function is discussed in detail in the next section. We also emphasize here that UDP is suitable for video applications not only because of their strict delay constraint, but also due to the QoS requirements. Compared with traditional elastic applications such as web browsing and electronic mail, real-time video applications are much more sensitive to delay but more tolerant to packet loss.

In addition, UDP has, although optionally, a checksum capability of verifying that the data have arrived intact. Usually, only intact packets are forwarded to the application layer. This makes sense for wired IP networks, where entire packets may be lost due to buffer overflow. In a wireless IP network, however, bit errors are present in the received packets. In this case, those packets with bit errors may still be useful to the application. For example, if the bit errors occur at the DCT coefficients part of the bitstream but the motion vectors are intact, these motion vectors can be used for either reconstructing the current packet or concealing neighboring packets. For this reason, modifications to the current UDP have been proposed and studied. One such modification is the UDP Lite [36]. Note that in IP-based wireless networks, video packets can also be transported utilizing the UDP protocol [37].

Real-time transport protocol (RTP) runs on top of UDP/IP for real-time applications. It provides end-to-end network transport functions suitable for applications transmitting real-time data, such as audio, video or simulation data, over multicast or unicast networks. It can be used for media on demand as well as interactive services. RTP consists of a data part and a control part. The latter is called RTCP (realtime transport control protocol) [38].

The data part of RTP is a thin protocol providing support for real-time applications, including timing reconstruction, loss detection, security, and content identification. RTCP provides support for data delivery in a manner scalable to large multicast networks. This support includes time stamping, sequence numbering, identification, as well as multicast-to-unicast translators. It offers QoS feedback from receivers to the multicast group as well as support for the synchronization of different media streams.

Another application-layer protocol for video applications is the real-time streaming protocol (RTSP), which establishes and controls one or more time-synchronized continuous media delivery with real-time constraints. An example of an RTSP application is the RealPlayer.

2.2.2   Packetization

At the sender of a video transmission system, video packets (also termed source packets) are generated by a video encoder. We refer to this stage as *source packetization*. In the application layer, source packets can be re-packetized into *intermediate packets* (e.g., for the reason of interleaving or FEC). After source packets pass through the network protocol stack (e.g., RTP/UDP/IP), they form *transport packets* to be sent over the network. The functionality of packetization that converts source packets into transport packets at the transport layer is referred to as *transport packetization*. Next, we introduce the general rules for source packetization in real-time video applications. More details on transport packetization are provided in the following chapters, since the transport packetization schemes are application dependent.

An important rule is that each video packet should be encoded independently without using predictive coding across packets. Due to this, each packet can be independently decoded, and the boundaries introduced by the packets can be used to limit the propagation of errors in the received bitstream. The RTP standard is based on the concept of application level framing introduced in [39]. The idea is that packetization should take into account natural boundaries in the data set by the application layer. For video applications, general rules for generating packetization schemes to be used with RTP can be found in [8, 40]. Examples of specific video formats can be found in [41] for MPEG-1 and MPEG-2 streams, in [42] for MPEG-4, in [43] for H.261, and in [8, 44] for H.264/AVC.

In setting the packet size, the fragmentation limit of intermediated nodes in IP networks should be taken into account. In the Internet, the maximum transfer unit (MTU), which is the maximum size of a packet that can be transmitted without being split/recombined at the transport and network layers, is equal to 1,500 bytes. For wireless networks, the MTU size is typically much smaller and 100 bytes are commonly assumed in most research activities, including JVT's wireless common conditions [8]. The size of source packets does not have to be constrained by the limitation imposed by RTP, since large or small source packets can be fragmented or aggregated at the IP layer to adapt to the appropriate RTP packet size.

However, if IP fragmentation is employed, protection efficiency will be reduced, because the techniques based on UEP, such as data partitioning, are obviously not available. Additionally, the creation of one large source packet at the encoder end enhances coding efficiency through predictive coding across macroblocks (MBs). Such efficiency will be greatly reduced if small source packets are created at the encoder end[1] and aggregated in the IP layer to adapt to the RTP size limitation[2]. For the above reasons, translation of one source packet directly into one RTP packet is usually preferred in most real-time video applications.

In addition, in order to use RTP/UDP/IP protocols, a typical RTP packet in IP networks requires a header of approximately 40 bytes per packet [45]. To minimize packetization overhead, the size of the payload data should be substantially larger than the header size. Therefore, considering the fragmentation limit of intermediated nodes in IP networks, the conceivable lower and upper bounds for the payload data per packet over the Internet may be one row of MBs and one entire coded frame, respectively.

Following the above rules, depending on the network infrastructure, each GOB (group of blocks) up to one video frame can be coded as one source packet, and every source packet is independently decoded. There is a clear engineering tradeoff with respect to the packet size, since smaller packets increase the packetization overhead while also increasing the robustness of the bitstream to channel errors [13]. For a given packetization scheme, the decoder clearly knows the packet boundaries, and lost motion information can be easily estimated through the received neighboring packets.

2.2.3   Network monitoring

The network monitoring techniques can be classified into active/passive, on-demand/ continuous, or centralized/distributed techniques according to various classification criteria [10].

In the Internet, CSI is usually estimated through a feedback channel (e.g., using RTCP). Specifically, from the headers of the transmitted and feedback packets, parameters, such as sequence number, time stamp, and packet size, can be collected at regular time intervals. Based on these parameters, CSI, such as the packet loss ratio, network RTT, retransmission timeout (RTO), and available network bandwidth, is estimated. One such example is provided in [11].

As to IP-based wireless networks, e.g., 3G wireless networks, one of the main services provided by the physical layer is the measurement of various radio link quantities, such as radio-link BER, channel signal-to-noise ratio (SNR), Doppler spectrum, and channel capacity. In

---

[1]Note that predictive coding across source packets is prohibited.
[2]Note that source packets much smaller than the RTP size limit do not have to be aggregated at the IP layer. But aggregation improves efficiency when a fixed RTP/UDP/IP header has to be attached to each RTP packet.

order to facilitate the efficient support of QoS for video applications, these measurements are reported to the upper layer for channel state estimation from the perspective of link layer or transport layer. Translation of physical-layer channel measurements to upper-layer channel parameters is necessary because the physical-layer channel models do not explicitly characterize a wireless channel in terms of the necessary QoS required by the video applications, such as data rate, packet loss probability, and delay. One scheme for the translation of physical-layer channel measurement into upper-layer channel parameters is presented in [37] to account for CSI estimation for video applications using UDP. Another scheme is the effective capacity model developed in [31].

2.2.4   Congestion control

Congestion control refers to the strategy employed to limit the sender's transmission rate to avoid exhaustion of network resources by a high volume of traffic. Congestion control is an indispensable component in most communication systems operating over best-effort networks. In order to transport media over the IP network efficiently, and in order to make fair use of the available resources, all IP service systems are expected to react to congestion by adapting their transmission rates. A good review of congestion control can be found in [1].

Congestion control techniques used for unicast video are usually sender-driven with two general categories: sender-based [46] and model-based [11]. Between the two, model-based TCP-friendly congestion control is usually recommended for video transmission over variable bit rate (VBR) channels due to its smooth transmission pattern. In addition, since the Internet today is dominated by TCP traffic, it is very important for multimedia streaming to be "TCP-friendly", meaning that a media flow must generate similar flow throughput as a TCP traffic under the same conditions, but with lower latency.

One example [11, 47, 48] of a stochastic TCP model that can be used to estimate the network throughput represents the throughput of a TCP sender as a function of a constant packet loss probability and RTT, and provides an upper bound of the sending rate. To further smooth out the sending rate fluctuations and thus facilitate video applications, Wu *et al.* [11] proposed a heuristic method adjusting the sending rate based on the estimated bandwidth and degree of network congestion. By using this method, the sending rate can be increased or decreased very smoothly according to the network-related information. Similar methods for smooth and fast rate adaptation congestion control can be found in [49, 50].

As for wireless networks, one congestion control model is presented in [51] to estimate the UDP throughput R_T by assuming a two-state Markov chain to describe the behavior of the success and failure of link-layer packets. In this model, R_T depends on the transport-channel bit rate or the probability of a successful UDP packet, given that the previous packet was received successfully or not, respectively. The relationship between packet transition probabilities and link-layer frame transition probabilities can be found in [37].

2.2.5   IP-based wireless network interface

Wireless IP networks are usually divided into two categories: indoor systems based on the IEEE 801.11 Wireless LANs (WLANs) and outdoor systems based on the emerging 3G and 4G wireless networks [52]. Here, we highlight some of the major differences of wireless IP networks from the Internet, in terms of the functionalities provided to support real-time video communications.

It is well known that the application-layer or transport-layer Automatic Repeat reQuest (ARQ) is usually not suitable for real-time video transmissions due to the unbounded delay it introduces. But the media access control (MAC) layer retransmission can react to changing channels faster and leads to smaller delays than its upper layers (MAC is a sublayer of the link layer). Therefore, the MAC layer retransmissions might be applicable for real-time video transmission. In IEEE 802.11 WLAN, the maximum number of MAC layer retransmissions, i.e., the retransmission limit, can be changed adaptively per packet to control the throughput, reliability, and delay tradeoffs. In addition, there are eight modulation and channel coding modes defined at the physical layer.

Retransmissions in the 3G and 4G systems, such as CDMA 2000, are more complicated. For such systems, after an IP packet is generated by passing through the transport and network layers, it is fragmented into packet data units (PDUs) at the link layer. The link layer protocol is called radio link control (RLC), which is a kind of selective repeat ARQ. RLC defines the frame length of PDU as 336 bits, which is the unit for retransmission. The key component here affecting performance is the behavior of the data link layer ARQ protocol. In addition, FEC and interleaving (e.g., RCPC/CRC is used to provide error protection and check) may be provided at the physical layer. The FEC rate and interleaving length may vary depending on the CSI, the available bandwidth, and the delay constraint. In addition, similar to IEEE 802.11, retransmissions can be realized at the MAC layer to better react to the changing channels. Thus, besides TCP, there are two levels of retransmission implemented at the link layer, which makes the performance evaluation of such systems very challenging [53].

As FEC and ARQ each has its own benefits with regard to error robustness and network traffic load, hybrid ARQ techniques, which can be used to describe any combined FEC and ARQ scheme, are widely used to achieve higher throughput in an error-prone wireless channel [24, 54]. In this scenario, FEC can be in the form of retransmission of the same packet or sending progressive parity packets in each subsequent transmission of the packet. The former is referred to as *chase combining* or type III hybrid ARQ and the latter as *incremental redundancy* or type II hybrid ARQ [55]. Incremental redundancy is especially appealing for efficient video transmissions over wireless IP networks. This is because FEC can be adapted to the time-varying channel conditions, and as a result, transmission of unnecessary redundancy bits can be avoided [56].

2.3   ERROR CONTROL TECHNIQUES

Video applications typically exhibit much higher latencies than voice applications. This is due to the strong interdependencies among video streams captured primarily by motion compensation. Temporally predictive video encoding is employed through motion compensation to achieve high compression by reducing temporal redundancies between neighboring frames. Furthermore, predictive coding is also employed to reduce spatial statistical redundancies, as exampled by differentially encoding the DC values of the DCT coefficients, the quantization parameters, and the motion vectors. Generally speaking, the higher the compression ratio, the greater the sensitivity of the video stream to channel errors. Errors, for example, caused by motion information loss can propagate temporally and spatially due to the use of predictive coding, until the prediction loop is restarted or synchronization is recovered [57]. Compressed video packets have different levels of sensitivity to packet loss leading to a different level of relative importance for each packet; thus, error control is a critical consideration in designing a video communication system.

Error control generally includes error-resilient source coding, FEC, retransmission, power adaptation, QoS differentiation in the network, and error concealment [1, 57]. The structure of the various error control components in a video transmission system is illustrated in Fig. 2.3. Each of the above-mentioned error control approaches is designed to deal with a lossy packet channel. Next we discuss each approach in some detail.

2.3.1   Error-resilient source coding

Error-resilient source coding refers to techniques of adding redundancy at the source coding level to prevent error propagation and limit the distortion caused by packet losses. There are various tools targeting different channel environments that can be used to achieve error resiliency. For example, channel errors have different forms in wireless channels (bit errors) and wired channels (packet losses). For wireless video, error-resilient source coding may include data partitioning, resynchronization, and reversible variable-length coding (RVLC) [13,58,59]. For packet-switched networks, error resiliency may include the selection of the encoding mode for each packet [58,60–62], the use of scalable video coding [63–65], and multiple description coding (MDC) [13,66,67]. In addition, packet dependency control has been recognized as a powerful tool to increase error robustness (the details are discussed in Chapter 4).

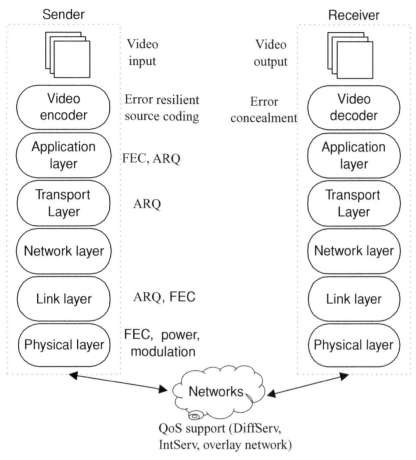

FIGURE 2.3: Illustration of error control components in a video transmission system.

2.3.2   Forward error correction

Although error-resilient source coding is a powerful tool to achieve robustness against channel errors, adaptation at the source cannot always overcome the large variations in channel conditions. It is also limited by the delay in the feedback channel, as well as the low accuracy and difficulty in estimating the time-varying channel conditions. Another way to deal with packet losses is to use error-correction techniques by adding channel coding redundancy. Two basic techniques are typically used: FEC and ARQ. Each has its own benefits in terms of error robustness and network traffic load [24, 54]. Of the two error-correction techniques, FEC is usually preferred for real-time video applications due to the strict delay requirements and semireliable nature of video streams [1, 13]. The FEC methods used depend on the requirements of the system and the nature of the channel. Details of FEC techniques are discussed in Chapter 5.

2.3.3   Retransmission

In addition to the introduced unbounded delay, ARQ may not be appropriate for multicast scenarios due to their inherent scalability problems [1, 13]. This is because retransmission typically benefits only a small portion of receivers while all others wait unproductively, resulting in poor throughput. However, the error detection/correction capability of FEC is limited due to the restrictions on the block size dictated by the application's delay constraints. FEC also incurs constant overhead even when there are no losses in the channel. In addition, the appropriate level of FEC heavily depends on the accurate estimation of the channel's behavior. ARQ, on the other hand, can automatically adapt to the channel loss characteristics by transmitting only as many redundant packets as are lost. Compared to FEC, ARQ can usually achieve a level closer to channel capacity. Of course, the tradeoff is that larger delays are introduced by ARQ. Thus, if the application has a relatively loose end-to-end delay constraint (e.g., on-demand video streaming which can tolerate relatively large delays due to a large receiver buffer and long delay for playback), ARQ may be a better choice. Even for real-time applications, delay constrained application-layer ARQ has been shown to be useful for some situations such as LAN, where RTT is relatively small [48, 54, 68, 69]. Various delay-constrained retransmission schemes for unicast and multicast have been discussed in [1]. The details of retransmission techniques are discussed in Chapter 5.

2.3.4   Transmission power control

In wireless channels, besides FEC, the characteristics of the wireless channel as seen by the video encoder can be changed by adjusting the transmitter power. Thus, prioritized transmission can be achieved through this transmitter power adjustment for each packet. Specifically, for a fixed transmission rate,[3] increasing the transmission power will increase the bit energy and consequently decrease the BER. Conversely, for a fixed transmission power, increasing the transmission rate will increase the BER but decrease the transmission delay needed for a given amount of data. Therefore in an energy-efficient wireless video transmission system, transmission power may need to be balanced against delay to achieve the best video quality at the receiver [6, 70].

2.3.5   Network QoS support

Error control can also be achieved through QoS support from the network, since video packets with different importance levels can be transmitted with different QoS guarantees. Architectures supporting QoS have been under discussion for over a decade. Two representative approaches

---

[3] Here we denote the transmission rate for the source in source bits per second, thus the corresponding BER is the source BER.

have been proposed in Internet Engineering Task Force (IETF) the integrated services (IntServ) with the resource reservation protocol (RSVP) [71,72] and the differentiated services (DiffServ or DS) [20,73]. IntServ supports QoS by reserving resources for individual flows in the network. The main disadvantage of IntServ is that it does not scale well to large networks with thousands of reserved flows, where each router must maintain per-flow state information.

In contrast, DiffServ supports QoS by allocating resources discriminatorily to aggregated traffic flows based on multiple service classes [20,21,73]. Basically, the sender assigns a priority tag (a DS byte) to each packet, indicating the QoS class to which the packet belongs. Upon arriving at a router, a packet is queued and routed based on its assigned class. Consequently, the DiffServ approach allows different QoS to cater to different classes of aggregated traffic flows. Typically, a packet assigned to a high-QoS class is less likely to be dropped or delayed at a router than a packet assigned to a low-QoS class. These per hop behaviors lead to an end-to-end statistical differentiation between QoS classes [20, 21].

2.3.6 Error concealment

Error concealment refers to post-processing techniques employed by the decoder. Since the human visual system can tolerate a certain degree of distortion in video signals, error concealment is a viable technique in handling packet loss. These methods can be broadly classified into spatial, temporal, and spatio-temporal domain approaches [74]. With spatial approaches, missing data are reconstructed using neighboring spatial information, whereas with temporal approaches, missing data are reconstructed from those in previous frames. Spatio-temporal error concealment approaches clearly represent the combination of the previous two approaches "Post-processing" and "Spatio-temporal" are widely used terms. Temporal concealment usually results in lower perceptual distortion than spatial concealment [60]. Most temporal concealment techniques use temporal replacement based on the motion information of neighboring MBs [13]. These techniques attempt to estimate the missing motion information from neighboring spatial regions in order to perform motion compensation to conceal errors. Several estimates have been studied, e.g., using the average, median, and maximum *a posteriori* (MAP) estimates, or a side-matching criterion [74, 75]. In [76], it was found that in most cases using the median estimate for motion compensation results in better subjective quality than the averaging technique. This is also the technique employed in the H.263 Test Model [77].

An example of a temporal replacement error concealment strategy, which is relatively simple to implement but similarly efficient to the ones in [4, 45], is shown in Fig. 2.4, where one packet consists of one row of MBs (11 MBs in this example). The motion vectors are spatially causal, i.e., the decoder will only use information from previously received packets in concealing a lost packet. When a packet is lost, the motion vector to be used for concealing an MB within the lost packet will be the median value of the three motion vectors at the top-left,

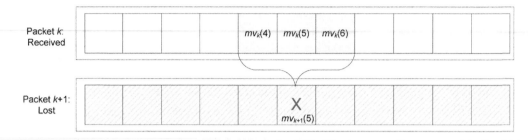

$$mv_{k+1}(5) = \text{median}(mv_k(4), mv_k(5), mv_k(6))$$

FIGURE 2.4: Illustration of an error concealment scheme.

top, and top-right MBs of the lost MB. If the previous packet is also lost, then the concealment motion vector is zero, i.e., the MB in the same spatial location in the previously reconstructed frame is used to conceal the current loss.

2.4   DISTORTION MEASUREMENT

For lossy compression, video quality is usually evaluated in terms of the difference between the original picture and the reconstructed one from the decoder. Generally speaking, when the quality of the reconstructed video is to be evaluated by a human, the properties of the human visual system should be taken into account in defining the distortion metric. The development of perceptually relevant image quality metrics is an active field of research [78–82]. The design of perceptual metrics for video is an even more challenging task than the design of still images due to the temporal dimension [83,84]. The mean squared error (MSE) and the peak signal-to-noise ratio (PSNR) are used for reporting results in most of the image and video processing literature. In the case of an image $f(m, n)$ of dimension $M \times N$ and its corresponding reconstruction $\hat{f}(m, n)$, the MSE is defined as

$$\text{MSE} = \frac{1}{MN} \sum_{m=1}^{M} \sum_{n=1}^{N} [f(m, n) - \hat{f}(m, n)]^2. \tag{2.1}$$

PSNR is inversely related to MSE, as $\text{PSNR(dB)} = 10 \log \frac{255^2}{\text{MSE}}$, where 255 is the maximum intensity value of an 8-bit per pixel image frame.

Although these metrics are defined without reference to any human perception model, it has been reported [16] that compression systems optimized for MSE performance also yield good perceptual quality. For a video sequence in the RGB format or the 4:4:4 YCbCr format, it is reasonable to calculate the PSNRs for each R/G/B or Y/Cb/Cr component and use their average value to denote the overall video quality, as these formats are targeted for high video

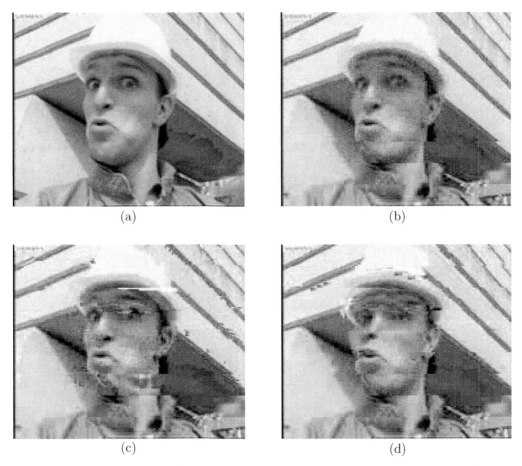

FIGURE 2.5: Illustration of the effect of random channel errors to a video stream compressed using the H.263 standard: (a) original frame; (b) reconstructed frame at the encoder; (c), (d) reconstructed frames at the decoder using two different runs of the simulation experiment (QCIF Foreman sequence, frame 90, coded with 96 kbps, frame rate 15 fps, and packet loss probability 10%).

quality applications. However, if the video is in the 4:2:2 or the 4:2:0 YCbCr format, it might be sufficient to merely use the PSNR for the Y component to describe the overall quality.

2.4.1   End-to-end distortion

In an error-prone channel, the reconstructed images at the decoder usually differ from those at the encoder due to the random packet losses, as shown in Fig. 2.5, in the form of an example. Even with the same channel characteristics, the reconstruction quality at the decoder may vary greatly based on the specific channel realization, as indicated in Figs. 2.5(c) and 2.5(d). In this case, the *expected* end-to-end distortion is utilized instead of the distortion between the original image and the reconstructed image at the encoder. That is, we measure the distortion between the original image, $f^{(n)}$, and the reconstructed image at the decoder, $\tilde{f}^{(n)}$, where the superscript (n) denotes the frame index. Specifically, by omitting the frame index, we calculate the end-to-end distortion for packet k as $E[D_k] = \frac{1}{K} \sum_{j=1}^{K} E[d_j(f_j, \tilde{f}_j)]$, where the subscript j denotes the pixels that belong to packet k, d_j is the distortion of the jth pixel, K is the number of pixels in packet k, and $E(\cdot)$ is the expectation calculated with respect to the probability of packet loss.

If MSE is employed, we have

$$E[D_k] = \frac{1}{K} \sum_{j=1}^{K} E[(f_j - \tilde{f}_j)^2]. \tag{2.2}$$

Alternatively, $E[D_k]$ can be written as

$$E[D_k] = (1 - \rho_k)E[D_{R,k}] + \rho_k E[D_{L,k}], \tag{2.3}$$

where $E[D_{R,k}]$ and $E[D_{L,k}]$ are respectively the expected distortion when the kth source packet is either received correctly or lost, and ρ_k is its loss probability. Due to channel losses and error propagation, the reference frames at the decoder and the encoder may not be the same. Thus, both $D_{L,k}$ and $D_{R,k}$ are random variables. $E[D_{R,k}]$ accounts for the distortion due to source coding as well as error propagation caused by interframe coding, while $E[D_{L,k}]$ accounts for the distortion due to concealment. Predictive coding and error concealment both introduce dependencies between packets. Because of these dependencies, the distortion for a given packet is a function of how other packets are encoded as well as their probability of loss. Accounting for these complex dependencies is what makes the calculation of the expected distortion a challenging problem.

For the concealment scheme mentioned in Section 2.3.6, the expected distortion for the kth packet, $E[D_k]$, can be described as

$$E[D_k] = (1 - \rho_k)E[D_{R,k}] + \rho_k(1 - \rho_{k-1})E[D_{C,k}] + \rho_k \rho_{k-1} E[D_{Z,k}], \tag{2.4}$$

where $E[D_{C,k}]$ and $E[D_{Z,k}]$ are the expected distortions after concealment when the previous packet is either received correctly or lost, respectively.

Methods for accurately calculating the expected distortion have recently been proposed [60–62]. With such approaches, it is possible, under certain conditions, to accurately compute the expected distortion with finite storage and computational complexity by using per-pixel accurate recursive calculations. For example, in [61], a powerful algorithm called recursive optimal per-pixel estimate (ROPE) is developed, which efficiently calculates the expected mean-squared error by recursively computing only the first and second moments of each pixel in a frame. Model-based distortion estimation methods have also been proposed (for example, [58,85,86]), which are useful when the computational complexity and storage capacity are limited.

As shown in Fig. 2.5, with the same channel characteristics, different simulation runs may produce large variations in the quality of the reconstructed images. A novel approach called variance-aware per-pixel optimal resource allocation (VAPOR) is proposed in [5] to deal with this. Besides the widely used expected distortion metric, the VAPOR approach aims to limit error propagation from random channel errors by accounting for both the expected value and the variance of the end-to-end distortion when allocating source and channel resources. By accounting for the variance of the distortion, this approach increases the reliability of the system by making it more likely that what the end-user sees, closely resembles the mean end-to-end distortion calculated at the transmitter.

Since the ROPE algorithm [61] is a widely used end-to-end distortion model and is also used in this monograph, we provide some additional details next.

2.4.2 ROPE algorithm

Assuming the MSE quality criterion, the overall expected distortion level of pixel j in frame n can be written as

$$E[d_j^{(n)}] = E[(f_j^{(n)} - \tilde{f}_j^{(n)})^2] = (f_j^{(n)})^2 - 2f_j^{(n)} E[\tilde{f}_j^{(n)}] + E[(\tilde{f}_j^{(n)})^2], \qquad (2.5)$$

where $f_j^{(n)}$ is the jth pixel of the nth original frame and $\tilde{f}_j^{(n)}$ the jth pixel of the nth expected reconstructed frame.

In the ROPE algorithm, the first- and second-order expected values of one pixel [the second and the third term on the right-hand side of Eq. (2.5)] are calculated recursively, as shown in [61]. Their calculation depends on the specific packetization scheme, the error concealment method used, and the pixel's prediction mode. If the error concealment scheme described in Section 2.3.6 is employed, the expected value of each pixel in the current frame only depends on the first- and second-order expected values of the pixels in the previous frame. Thus for calculating expected values for each pixel, only the first- and second-order pixel values of one frame need to be stored at any given time.

Here we emphasize that the inter frame error propagation due to channel errors has been captured in this distortion characterization. This is because in order to calculate the expected distortion of the current frame, all that needs to be determined is the first- and second-order information of each pixel's values in the previous frame due to the nature of motion compensation. For the study of the accuracy of this distortion measurement, readers are referred to the original paper [61] and the updated work [87].

CHAPTER 3

Joint Source-Channel Coding

As mentioned in Chapter 2, lossy compression is generally required for video transmission applications. It is well known that the theoretical bounds for lossless compression are the entropy of the source. In the same way entropy determines the lowest possible rate for lossless compression, rate-distortion (RD) theory [18, 19] addresses the same question for lossy compression. In this chapter, we begin with a basic description of RD theory, and then describe the reason of why we need joint source-channel coding, as well as the specific task and formulation of joint source-channel coding in video communication applications. After that, we discuss the general tools utilized for solving the resulting optimization problem in JSCC.

3.1 RATE-DISTORTION (RD) THEORY

A high-level block diagram of a communication system is shown in Fig. 3.2. In this diagram, X and $\hat{X}$ represent, respectively, the source and reconstructed signals, X_s and $\hat{X}_s$ the source encoder output and the source decoder input, and X_c and $\hat{X}_c$ the channel encoder output and the channel decoder input.

Rate distortion theory is concerned with the tradeoffs between distortion or fidelity and rate in lossy source compression schemes. In other words, there is either no transmission of the compressed source involved, or the channel can be considered to be lossless, as seen by the source encoder (we will further discuss this in Section 3.3). This situation is depicted in Fig. 3.1, where $X_c = \hat{X}_c$ and $X_s = \hat{X}_s$ have been used.

Rate is defined as the average number of bits used to represent each sample value. The general form of the source distortion D can be written as

$$D = \sum_{x_i, \hat{x}_j} d(x_i, \hat{x}_j) p(x_i) p(\hat{x}_j | x_i), \qquad (3.1)$$

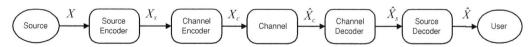

FIGURE 3.1: Block diagram of a communication system.

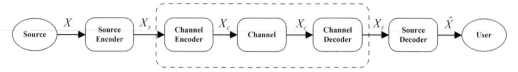

FIGURE 3.2: Block diagram of a communication system (to illustrate RD theory).

where $d(x_i, \hat{x}_j)$ is the distortion metric, $p(x_i)$ is the source density, and $p(\hat{x}_j|x_i)$ is the conditional probability of the reconstructed value $\hat{x}_j \in \hat{X}$, given knowledge of the input to the encoder $x_i \in X$. Since the source probabilities are solely determined by the source and the distortion metric is decided ahead of time depending on the application (see also Section 2.4), the distortion is a function only of the conditional probabilities $p(\hat{x}_j|x_i)$, which represent a description of the compression scheme. We can therefore write the constraint as the distortion D is less than some value D^*, requiring that the conditional probabilities for the compression scheme belong to a set of conditional probabilities Π that have the property that

$$\Pi_D = \left\{ \{p(\hat{x}_j|x_i)\} : D(\{p(\hat{x}_j|x_i)\}) \leq D^* \right\}. \tag{3.2}$$

The centerpiece of RD theory is the RD function $R(D)$, which provides the theoretical information bound on the rate necessary to represent a certain source with a given average distortion. It is given by [19]

$$R(D) = \min_{\{p(\hat{x}_j|x_i)\} \in \Pi_D} I(X; \hat{X}), \tag{3.3}$$

where $I(X; \hat{X})$ is the average mutual information between X and $\hat{X}$, defined as

$$I(X; \hat{X}) = \sum_{x_i, \hat{x}_j} p(x_i, \hat{x}_j) \frac{p(x_i, \hat{x}_j)}{p(x_i)p(\hat{x}_j)}. \tag{3.4}$$

The average mutual information between the source output and the reconstruction is a measure of the information conveyed by the reconstruction about the source output (entropy of the source minus the uncertainty that remains about the source output after the reconstructed value has been received). Since the desired distortion has been specified, i.e., the minimization in (3.3) is over the set of conditional probabilities which satisfy the distortion constraint, the minimization of the average mutual information does represent the rate.

The RD function $R(D)$ has some nice properties, e.g., it is a nonincreasing convex function of D. It can be used to find the minimum bit rate for a certain source with a given distortion constraint, as expressed by (3.3). Conversely, it can also be used to determine the information theoretical bounds for the average distortion subject to a source rate constraint,

R^*, via the distortion-rate function, which is the dual of (3.3), defined as

$$D(R) = \min_{\{p(\hat{x}_j|x_i)\} \in \Pi_R} D(X; \hat{X}),$$ (3.5)

where R is the average source rate and

$$\Pi_R = \{\{p(\hat{x}_j|x_i)\} : R(\{p(\hat{x}_j|x_i)\}) \leq R^*\}.$$ (3.6)

Note that the $D(R)$ function may be more widely applicable in practical image/video communication systems, since, as a practical matter, the aim is usually to deliver the best quality image/video subject to a certain channel bandwidth. RD theory is of fundamental importance in that it conceptually provides the information bounds for lossy data compression. However, the proof of the RD function is an existence proof [88]. That is, it only tells us that $R(D)$ exists but not how to achieve it in general (except for a few special cases such as the case of a Gaussian source). In addition, as already mentioned, RD theory only addresses source coding. In a communication system where the channel may introduce errors, we need ideal channel coding to ensure that $X_s = \hat{X}_s$ in order for (3.5) to be valid, where $D(R)$ is interpreted as the end-to-end distortion and R^* as the channel capacity. Thus, there are several practical issues that prevent us from obtaining the RD function or designing a system to achieve it. In the following two sections, we discuss these issues.

3.2   OPERATIONAL RATE-DISTORTION (ORD) THEORY

From (3.3), we see that $R(D)$ is evaluated based on information which may not always be available in practice. First, the source density required to evaluate the distortion, which in turn is needed in (3.3) or (3.5), is not typically available for any given video segment. If a specific model is to be used, in general it does not capture the spatiotemporal dynamics of a given video well. Second, the minimization in (3.3) or (3.5) is over all possible coders or conditional distributions $p(\hat{x}_j|x_i)$ which belong to Π_D or Π_R, respectively. In a practical coding environment, however, we want to use RD theory to allocate resources to a specific video segment using a specific coder. In this case, we usually only have a finite admissible set for $p(\hat{x}_j|x_i)$, as defined by the finite set of encoding modes (e.g., intra or interencoding, quantization step size, etc.). In addition, it is usually difficult or simply impossible to find closed-form expressions for the $R(D)$ or $D(R)$ functions for general sources. For these reasons, one has to resort to numerical algorithms to specify the operational RD (ORD) function.

Let S be the set of admissible source coding parameters or modes. Then since the structure of the coder has been determined, each of the parameter choices will lead to a pair of rate and distortion values. Such a pair of operational points is indicated by an ($\times$) in the RD domain, as shown in Fig. 3.3. The lower bound of all these rate-distortion pairs is referred to as the ORD function, which is shown by a dashed line in the same figure. The set of source coding parameters that results in the ORD function can be formally defined as

$$\mathcal{U}_{\text{ORDF}} = \{q : q \in \mathcal{S}, R(q) \geq R(p) \Rightarrow D(q) < D(p), \forall p \in \mathcal{S}\}. \tag{3.7}$$

Clearly the ORD function should be lower bounded by the RD function, because the former is obtained using a specific coder and the finite set $\mathcal{S}$, which is a subset of the set of source codes that achieve all conditional distributions $p(\hat{x}_j|x_i)$.

According to (3.7), the ORD function is a strictly decreasing function. Although the RD function is a convex function, it is not necessary that an ORD function is also convex, as shown in Fig. 3.3. However, if the size of the admissible set $\mathcal{S}$ is sufficiently large, the ORD function will be closely approximated by the convex hull of all the operating points, which is a convex function. As is well known, if the objective function in an optimization problem is convex, a local minimum is also a global minimum [89]. The resulting convex hull approximation solution is very close to the optimal solution, if the set $\mathcal{S}$ is sufficiently large, which is usually the case in video compression.

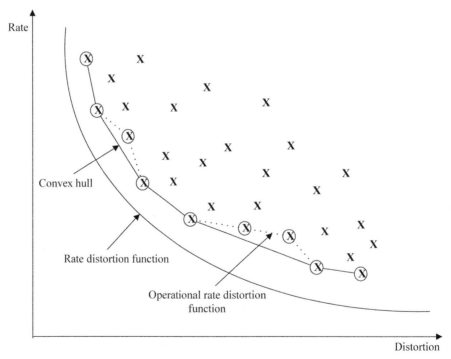

FIGURE 3.3: Illustration of the operational rate-distortion function.

3.3   ON JOINT SOURCE-CHANNEL CODING

RD theory, as already mentioned, only addresses source coding, since the search over the conditional probabilities $p(\hat{x}_j|x_i)$ is not coupled with the channel. In a communication system, as shown in Fig. 3.2, the RD function can be obtained if we assume that the source encoder output X_s is identical to the source decoder input $\hat{X}_s$. This can be achieved, according to the channel coding theory, if the source rate generated by X_s is less than the channel capacity C, which is defined as $C = \max_{p(x_c)} I(X_c; \hat{X}_c)$, where $x_c \in X_c$ and the maximum is taken over all possible channel input distributions $p(x_c)$. What channel coding does in essence is to find the best way to map X_s to X_c such that the distribution of X_c leads to a large minimum distance between codewords with the largest mutual information between the channel input X_c and channel output $\hat{X}_c$.

Combining the source coding and channel encoding blocks, rate-distortion theory and channel coding theory basically define the maximum achievable fidelity of a given source transmitted over an error-prone channel. That is, by letting the source rate constraint in Eq. (3.6) be the channel capacity, i.e., $R^* = C$, Eq. (3.5) can be used to determine the minimum achievable end-to-end distortion for a given source transmitted over an error-prone channel with capacity C. This is shown in Fig. 3.1, where the channel encoder and decoder can be combined with the channel, resulting in a lossless channel as seen by the source encoder. It suggests that source coding and channel coding can be designed separately in order to achieve the rate-distortion bound in Eq. (3.3), which is the basis of Shannon's separation theory [19].

However, such an assumption requires an ideal channel coding scheme, such that error-free transmission of the source output over a noisy channel with source bit rate $R(D)$ less than the channel capacity C, i.e., $R(D) < C$, can be guaranteed. Such ideal channel coding generally requires infinite length code words, which can only be realized without complexity and delay constraints, both of which are important factors in practical real-time systems. Due to these constraints, most practical channel coding schemes cannot achieve channel capacity and do not provide an idealized error-free communication path between the source and the destination. For this reason, the overall distortion between X and $\hat{X}_c$ consists of both source distortion and channel distortion. Minimizing the total distortion usually requires joint design of the source and channel encoders, as shown in Fig. 3.4, which is referred to as *joint source–channel coding*.

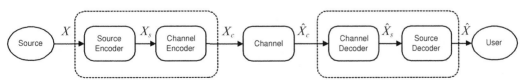

FIGURE 3.4: Block diagram of a communication system (to illustrate JSCC).

At the receiver side, gains may be realized by jointly designing the channel and source decoders, which is referred to as *joint source-channel decoding*. In using joint source-channel decoding, the channel decoder does not make *hard* decisions on the output $\hat{X}_s$. Instead, the decoder makes *soft* decisions in order to allow the source decoder to make use of information such as the signal-to-noise ratio (SNR) of the corrupted code. Alternatively, such soft decisions can be regarded as hard decisions plus a confidence measure. Soft-decision processing used in joint source-channel decoding can usually help improve the coding gain by about 2 dB compared to hard-decision processing [90]. In this monograph, we focus only on joint source-channel coding.

As for the joint design of source and channel coding, we extend the set of coding parameters from the source coding set S to the set U which also includes the channel coding parameters. Then for each choice of parameters in the set U, we plot its associated rate-distortion point in the rate-distortion map, as shown in Fig. 3.5. As we can see, all operational points are now shifted upwards and to the right, with reference to the points plotted in Fig. 3.3, where an error-free channel resulting from idealized channel coding is assumed. That is, in a practical system, we have a higher lower bound for all the generated rate-distortion pairs, which is shown as the dashdot curve in Fig. 3.5, compared to the dotted curve which is the ORD function

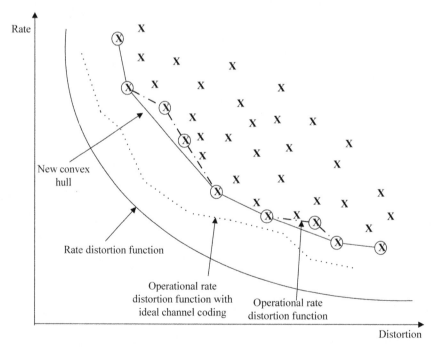

FIGURE 3.5: Illustration of operational rate-distortion function in a practical system.

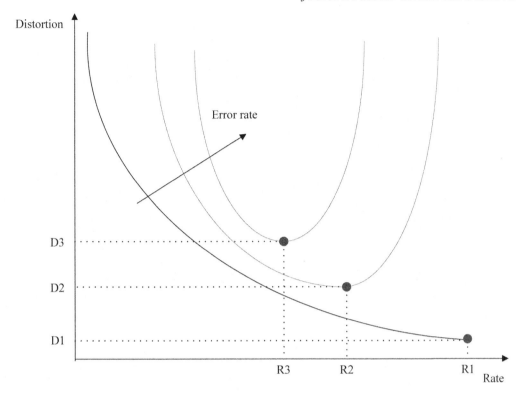

FIGURE 3.6: Rate-distortion function for various channel conditions.

with ideal channel coding. In the following we take another look at the rate-distortion tradeoff which also addresses the relative shift between the two ORD curves in Fig. 3.5.

In Fig. 3.6, we plot several ORD functions for a practical system at different channel error rates. As illustrated in the figure, when the channel is error free, increased bit rate leads to decreased distortion, as in standard rate-distortion theory. This is illustrated by the lowest curve in the figure, in which the lowest distortion is obtained by utilizing the largest available source bit rate, represented by the point (R1, D1). However, when channel errors are present, this trend may not hold, since the overall distortion now (the vertical axis in Fig. 3.6) consists of both source and channel distortion. Assuming that the transmission rate is fixed, as more bits are allocated to source coding (the horizontal axis in Fig. 3.6), fewer bits will be left for channel coding, which leads to less protection and higher channel distortion. As shown in Fig. 3.6, an optimal point exists for which the available fixed transmission rate bits optimally allocated between the source and the channel. Note that different channel-error rates result in different optimal allocations. This is indicated by points (R2, D2) and (R3, D3) on the two curves with different channel-error rates.

3.4   PROBLEM FORMULATION

The tradeoff between source coding and channel coding has been studied from a theoretical standpoint based on the use of vector quantizers in [91, 92]. In general, JSCC is accomplished by designing the quantizer and entropy coder jointly for given channel characteristics, as in [91, 93]. There are a substantial number of research results in this area. Interested readers can refer to [94] for a comprehensive review on this topic. In this section, we focus on the specific application of JSCC in image and video communications, where JSCC usually faces three tasks: finding an optimal bit allocation between source coding and channel coding for given channel loss characteristics, designing the source coding to achieve the target source rate, and designing the channel coding to achieve the required robustness [1, 62]. These tasks, although stated separately, are interrelated, forming the backbone of the integrated nature of JSCC.

As we can see, JSCC for image/video transmission applications is a typical optimal bit allocation problem, which has been studied for a long time. Next, we will first present the general problem formulation for optimal bit allocation, and then extend it to a resource-distortion optimization framework, which is suitable for systems that have additional resource constraints such as power or cost. Following that, we will briefly discuss the general approaches for the solution of the resulting problems.

3.4.1   Optimal bit allocation

In a typical formal approach of problem solving, there are four distinctive steps. First, an appropriate system performance evaluation metric should be selected. Second, the constraints need to be specified. Third, a model of the relationship between the system performance metric and the set of adaptation parameters needs to be established. Finally, the best combination of adaptation parameters that maximize the system performance while meeting the required constraints needs to be identified. Keeping those four steps in mind, we next present the formal approach to formulate the joint source–channel coding problem and provide solution approaches to such a problem.

As discussed in Section 2.4 a commonly used criterion for the evaluation of system performance in video transmission applications is the expected distortion. As shown in (2.3), in calculating the expected distortion for each source packet, the two distortion terms, $E[D_{R,k}]$ and $E[D_{L,k}]$, and the loss probability for the source packet ρ_k need to be determined. The two distortion terms depend on the source coding parameters such as quantization stepsize and prediction mode, as well as the error concealment schemes used at the decoder. The relationship between the source packet loss probability and channel characteristics depends on the specific packetization scheme, the channel model, and the adaptation parameters chosen.

Let $\mathcal{S}$ be the set of source coding parameters and $\mathcal{C}$ the set of the channel coding parameter. Let $s = \{s_1, ..., s_M\} \in \mathcal{S}^M$ and $c = \{c_1, ..., c_M\} \in \mathcal{C}^M$ denote, respectively, the vector of source coding parameters and channel coding parameters for the M packets in one video frame or a group of frames. The general formulation of the optimal bit allocation problem then is to minimize the total expected distortion, i.e., provide the best video delivery quality, for the frame(s), given the corresponding bit rate constraint [95], i.e.,

$$\min_{s \in \mathcal{S}^M, c \in \mathcal{C}^M} E[D(s, c)]$$
$$\text{s.t.} \quad R(s, c) \leq R_0, \tag{3.8}$$

where $E[D]$ is the total expected distortion, $R(s, c)$ the total number of bits used for both source and channel coding, and R_0 is the bit rate constraint for the frame(s).

Given a channel transmission rate R_T, the bit rate constraint can be converted to a transmission delay constraint $T(s, c) \leq T_0$, where $T(s, c) = R(s, c)/R_T$ is the transmission delay and $T_0 = R_0/R_T$ is the delay constraint. For a communication channel with a varying transmission rate as a function of the available resources, the delay constraint becomes more general and flexible. The transmission delay constraint T_0 is usually implicitly determined by the application based on the estimated channel throughput. For applications with very short end-to-end delay constraints, such as videoconferencing, the initial setup time T_{max} is usually very small (i.e., there is little additional buffering time at the receiver). For such applications, T_0 is very strict and usually equal to the duration of one frame, $1/F$, where F is the frame rate. However, for applications that have relatively loose end-to-end delay such as on-demand video streaming, T_{max} is generally much longer than $1/F$, due to the additional buffering at the receiver. In this case, T_0 is not that strict and usually varies around $1/F$ as a function of the video content (e.g., complicated frames usually require more bits for their encoding, or, alternatively, higher delay in transmission), and the dynamics of the encoder buffer and the playback buffer at the receiver. The determination of T_0 for a video group is achieved by rate control. Because rate control is usually separately designed from error control, we do not address this former component in this monograph. However, we recognize that rate control is a very important component in the overall end-system design.

The JSCC problem formulation (3.8) is general since both the source coding and the channel coding can take a variety of forms, depending on the specific application. Source coding parameters, for example, can be represented by the quantizer step sizes, mode selection [4,62,96], packet sizes [27], intra-MB refreshment rate [85], and entropy coding mechanisms [91]. As to channel coding, when FEC is utilized, the packet loss probability becomes a function of the FEC chosen. The details of this model will depend on how transport packets are formed from the available video packets [96]. Joint source coding and FEC have been extensively studied in the literature [45,63,97–99]. Such studies focus on the determination of the optimal bit allocation between source coding and FEC. In addition to FEC, retransmission-based error control may be used in the form of ARQ protocols. In this case, the decision whether to retransmit a packet or to send a new one forms another channel coding parameter, which also affects the probability of loss as well as the transmission delay. Note that since video packets are usually of different importance, the optimal bit allocation will result in an UEP across video packets.

When considering the transmission of video over a network, a more general joint source-channel coding scheme may cover the choice of modulation and demodulation [100], power adaptation [4], packet scheduling [101], and data rate adaptation [101]. These adaptation components can all be regarded as channel coding parameters. In those cases, additional constraints may apply, e.g., a transmitter energy constraint or a computational power constraint for a mobile device in a wireless video transmission application. In a DiffServ system, the user may have a cost constraint on what level of QoS of transmission can be achieved [3]. In light of these, we can in general refer to the bandwidth (bit rate) and the other constraints, such as energy and cost, as network resources constraints. Thus, we provide next the general resource-distortion optimization framework for video transmission applications.

3.4.2 Resource-distortion optimization framework

Now let $\mathcal{C}$ be the set of the network adaptation parameters. The formulation of a resource-distortion optimized joint source-network coding can be formally written as [95],

$$
\min_{s \in \mathcal{S}^M, c \in \mathcal{C}^M} E[D(s, c)]
$$
$$
\text{s.t.} \quad C(s, c) \leq C_0
$$
$$
T(s, c) \leq T_0, \tag{3.9}
$$

where C_0 and T_0 are the cost and transmission delay constraints, respectively, for the frame(s) under consideration. Unlike the rate constraint, the cost constraint C_0 is usually explicitly determined by the specific application. For example, for the application of transmitting video from a mobile device to the base station, the energy constraint is dictated by the battery life of the mobile device. For the application of DiffServ-based video transmission, the cost constraint comes from the negotiation of the user and the internet service provider (ISP) through the service level agreement (SLA).

The resource-distortion optimization framework is general for the joint design of error-resilient source coding, error concealment, and cross-layer resource allocation[1] and is flexible

---

[1]The cross-layer resource allocation we refer to is limited to those layers whose resource allocation parameters can be specified and controlled at the application layer. This usually requires new protocols that enable the application layer to specify those adaptation components at the lower layers.

to be applied to different applications. By solving the problem in (3.9) and selecting the source and channel coding parameters within their sets, we can obtain the optimal tradeoff among all those adaptation components. For example, by using this framework, an RD-optimized hybrid error control scheme has been presented in [102], which results in the optimal allocation of bits among source, FEC, and ARQ. In [6], joint source-channel coding and power adaptation has been studied for wireless video transmissions. In this chapter, we focus on the presentations of the optimization framework itself. Chapters 6 and 7 are devoted to the special cases of cross-layer resource allocation in different network infrastructures.

As mentioned in Chapter 1, our primary focus here is on conversational video applications and certain streaming video applications, since these type of applications require a close interaction between source and channel coding. Therefore, this resource-distortion optimization framework is mainly applicable for single users in a point-to-point unicast scenario. Study of single-user unicast case provides insight into a single component of more complicated scenarios involving multiple users and links. Efficient resource allocation for multiuser and multicast video transmission systems is another active research area with unique challenges. Issues such as fairness as well as distributed versus centralized resource allocation must be addressed in this setting.

3.5   SOLUTION ALGORITHMS

In this section, we briefly discuss the solution algorithms for the optimization problem (3.8). Although there are a large number of algorithms for solving constrained optimization problems, we focus on the Lagrangian relaxation method or the Lagrange multiplier method, as it represents an efficient method for solving constrained optimization problems where the objective and the constraint functions are both convex. For this reason, the Lagrangian method has been widely used in image/video compression and transmission applications, and also it is used throughout this monograph.

The key idea of Lagrangian relaxation is to introduce a variable called a Lagrange multiplier that controls the weights of the constraint when added as a penalty to the objective function. The constrained problem is then converted into an unconstrained problem as

$$\{s^*(\lambda), c^*(\lambda)\} = \arg \min_{s \in \mathcal{S}^M, c \in \mathcal{C}^M} \{E[D(s, c)] + \lambda R(s, c)\}, \qquad (3.10)$$

where $\lambda \geq 0$ is the Lagrange multiplier. The solution of (3.8) can be obtained by solving (3.10) with the appropriate choice of the Lagrange multiplier so that the bit rate constraint $R(s^*(\lambda), c^*(\lambda)) \leq R_0$ is satisfied. For the proof of this theory, readers can refer to [15,17]. The solution resulting from solving (3.10) is always on the convex hull of the ORD function for (3.8), which we will further discuss in the next section. In practice, due to the finite set of source and channel coding parameters, the objective function and the constraints are not continuous, thus the constraint may not be met with equality. In this case, the solution obtained by solving (3.10) will be the convex hull approximation solution to (3.8).

The solution process of the original constrained optimization problem can then be divided into two stages. Stage 1 requires the solution of the minimization problem of (3.10) for a given Lagrange multiplier. Stage 2 is to search for the appropriate Lagrange multiplier so that the constraint is met. The difficulty in solving the resulting unconstrained minimization problem depends on the complexity of the interpacket dependencies. Depending on the nature of such dependencies, an iterative descent algorithm based on the method of alternating variables for multivariate minimization [103] or a dynamic programming algorithm [104] can be employed to efficiently solve the minimization problem. Next we briefly describe them.

3.5.1   Lagrange multiplier method

For notational simplicity, let $\mathcal{U}$ denote the coding parameter set $\mathcal{S} \times \mathcal{C}$ and let $u \in \mathcal{U}$ be an element of this set, a coding parameter including both source and channel coding parameters. Assuming that $u^*(\lambda)$ is an optimal solution to (3.10), we then have

$$D(u^*(\lambda)) + \lambda R(u^*(\lambda)) \leq D(u) + \lambda R(u), \forall u \in \mathcal{U}, \tag{3.11}$$

which can be rewritten by rearranging terms as

$$R(u) \geq -\frac{1}{\lambda} D(u) + \{R(u^*(\lambda)) + \frac{1}{\lambda} D(u^*(\lambda))\}. \tag{3.12}$$

This means that all rate-distortion pairs $\{R(u), D(u)\}$ must lie on the upper right side in the rate-distortion plane of the line defined by (3.12) using an equal sign, as shown in Fig. 3.7. Note that $\{R(u^*(\lambda)), D(u^*(\lambda))\}$ is on the line and the line has slope $-\frac{1}{\lambda}$. For a different value of λ, the line with slope of $-\frac{1}{\lambda}$ will meet at least one optimal solution $\{R(u^*(\lambda)), D(u^*(\lambda))\}$. Increasing the value of the Lagrange multiplier from λ to λ_2 makes the line more horizontal, which leads to a smaller $R(u^*(\lambda_2))$ $(\leq R(u^*(\lambda)))$ and a larger $D(u^*(\lambda_2))$ $(\geq D(u^*(\lambda)))$, and vice versa. Thus, $R(u^*(\lambda))$ is a nonincreasing function of λ and $D(u^*(\lambda))$ is a nondecreasing function of λ. By sweeping λ from zero to infinity, the line will form a convex hull of all the RD pairs resulting from the coding parameter set $\mathcal{U}$, and all the optimal points on the convex hull can be reached, as shown in Fig. 3.7. Note that not all optimal points on the ORD function belong to the convex hull. These points are inaccessible by the Lagrange multiplier method, which represents a disadvantage of this method.

Having established that $R(u^*(\lambda))$ is a nonincreasing function of λ and $D(u^*(\lambda))$ is a nondecreasing function of λ, we can utilize the efficient bisection method shown in Fig. 3.8 to iteratively search for the appropriate λ so that the rate constraint in (3.8) is met.

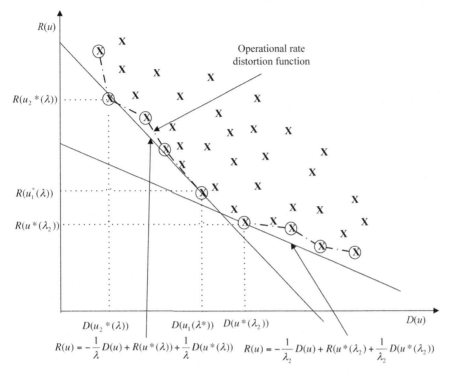

FIGURE 3.7: Illustration of the Lagrange multiplier method ($\lambda_2 > \lambda$).

Step 1: Let $\lambda^r = 0$ and λ^l be a relatively large value such that $R(u(\lambda^l)) \leq R_0 \leq R(u(\lambda^r))$.
Step 2: Let $\lambda^m = (\lambda^l + \lambda^r)/2$.
Step 3: If $R_0 - \delta < R(u(\lambda^m)) < R_0$ (with $\delta > 0$ a relatively small number), then the solution is $\lambda^* = \lambda^m$. Otherwise,
Step 4: if $R(u(\lambda^m)) \geq R_0$, then let $\lambda^l = \lambda^m$. Otherwise, let $\lambda^r = \lambda^m$. Go back to **Step 2.**

FIGURE 3.8: Bi-section method for Lagrange multiplier search.

Note that a limit on the number of loops is usually necessary, in case the condition in Step 3 cannot be met. This may result from the fact that δ is set at a level that is too small or the fact that optimal point, which is on the ORD function, is far away from the convex hull. This can be avoided if the size of the set $\mathcal{U}$ is large enough, because the continuous RD function is a nonincreasing convex function. Based on this property, some very fast algorithms have been proposed to search for the appropriate Lagrange multiplier. Readers can refer to [17] for details.

3.5.2 Dynamic programming

Dynamic programming (DP) is an approach developed to solve sequential or multistage decision problems. It is a widely employed tool in the field of optimization and has been studied extensively [104]. Our aim here is to describe the basic idea of DP and show how it can be used to solve the optimization problem (3.10) for a given Lagrange multiplier.

The essence of DP is that it solves a multivariable problem by solving a series of single variable problems. The key feature of a problem solvable by DP is that such a problem can be divided into *stages* with a *decision* made at each stage. Each stage has an associated *return* which is what we want to maximize (or minimize depending on how the return is defined). Each stage also has an associated *state* which has to be defined in such a way that it completely describes the system, which establishes the fact that the optimal decisions at the remaining stages only depend on the current state but not on the history before reaching this state. For a given state at one stage, the decision transforms the system state into another state at the next stage. Thus the system behavior moving from one stage to another forms a Markovian process, based on which there exists a recursive relationship of the optimal decisions for one stage given that the previous stage has already been solved. This is called the *Principle of Optimality*, which is the core of the DP algorithm. Thus, given the state at the initial stage and the ending state at the last stage, the optimization can be recursively solved. Next we present the mathematical expressions of the DP algorithm.

Assume that our objective is to maximize the total returns for $N+1$ stages. Let $r_k(s_k, x_k)$ be the return at the kth stage given state s_k and decision x_k at this stage, i.e., we want to obtain

$$\max_{\{x_0, \dots, x_N\} \in \mathcal{X}_0 \times \cdots \times \mathcal{X}_N} \sum_{k=0}^{N} r_k(s_k, x_k), \qquad (3.13)$$

where $\mathcal{X}_k$ defines the set of the decisions at the kth stage. Let $J_k(s_k) = \max \sum_{i=0}^{k} r_i(s_i, x_i)$ be the optimal total return starting from stage 0 until stage k ending at stage s_k. Then we have the following:

$$J_0(x_0) = r_0(s_0)$$
$$J_k(s_k) = \max_{x_k \in \mathcal{X}_k} \{r_k(s_k, x_k) + J_{k-1}(s_{k-1})\}, \quad 1 \leq k \leq N. \qquad (3.14)$$

Given the state of the initial stage s_0 and the ending state s_N, we can obtain $J_N(s_N) = \max \sum_{k=0}^{N} r_k(s_k, x_k)$, which is identical to (3.13), by recursively solving (3.14). At each stage when solving (3.14), the search space is $|\mathcal{X}_k| \cdot |\mathcal{S}_k|$, where $\mathcal{S}_k$ is the set of states and $|\cdot|$ denotes the cardinality of a set.

The difficulty in using DP lies in the identification of the appropriate states and stages for a given problem, as sequential stages may not be a feature of a given problem. We will not expand on this topic here; interested readers can refer to [104] or similar references for a comprehensive coverage of this topic.

The advantage of the DP algorithm may not be clear from the previous brief discussion of it. Next, we discuss the application of the DP algorithm to the problem at hand in (3.10), through which we illustrate how the structure of the problem lends itself to the use of DP and what the computational savings are.

Following the discussion above, we first need to identify the stages and states in (3.10). Let us assume that we perform optimal resource allocation among M packets that belong to a frame or a group of frames, and let $g_k = D_k + \lambda R_k$ be the objective function for packet k. As discussed in Chapter 2, for most applications where the MSE is used as a distortion metric, the distortion is additive and, of course, the rate is always additive. For this reason, the objective function is also additive[2]. Then the minimization problem can be rewritten as follows:

$$\boldsymbol{u}^*(\lambda) = arg \min_{\boldsymbol{u} \in \mathcal{U}^M} \sum_{k=1}^{M} g_k(u_1, \ldots, u_M), \qquad (3.15)$$

where we use the simplified notation of Section 3.5.1, and denote by u_k the coding parameter for the kth packet and by $\boldsymbol{u} = \{u_1, \ldots, u_M\} \in \mathcal{U}^M$ the coding parameter vector for the M packets. It is a natural choice to have each stage representing the coding of a packet in this case. As mentioned in Chapter 2, in video coding, the rate-distortion tradeoff of one MB or one packet typically depends on the coding parameter selected for its neighboring MBs or packets, due to the use of spatially predictive coding (e.g., the DC values of the DCT coefficients, the quantization parameters, and the motion vectors are usually differentially encoded), temporally predictive coding (motion compensation), and error concealment. Thus, it is reasonable to assume that the objective function or the return for each stage g_k only depends on the decision parameters in its neighborhood, that is

$$g_k = g_k(u_{k-a}, \ldots, u_{k+b}),$$

where a and b are both nonnegative integers. Then we can write (3.15) as

$$\boldsymbol{u}^*(\lambda) = arg \min_{\boldsymbol{u} \in \mathcal{U}^M} \sum_{k=1}^{M} g_k(u_{k-a}, \ldots, u_{k+b}). \qquad (3.16)$$

---

[2]If the objective function is not additive, we may not be able to use DP techniques to solve the optimization problem, since the problem may not be decoupled into a series of sequential problems.

Let $J_k(u_{k-a+1}, \ldots, u_{k+b}) = \min_{\{u_0, \ldots, u_{k-a}\} \in \mathcal{U}^{k-a+1}} \sum_{i=0}^{k} g_i(u_{i-a}, \ldots, u_{i+b})$, for $1 \leq k \leq M$. After some manipulations [17], we have

$$J_k(u_{k-a+1}, \ldots, u_{k+b}) = \min_{u_{k-a} \in \mathcal{U}} \left\{ g_k(u_{k-a}, \ldots, u_{k+b}) + J_{k-1}(u_{k-a}, \ldots, u_{k+b-1}) \right\}. \qquad (3.17)$$

Based on the discussion of the DP algorithm, we can easily identify the states and decisions of this system. Recall that the state completely characterizes the system such that it leads to a Markov process when the system moves from stage to stage. Thus we can define at stage k the state as $s_k = \{u_{k-a+1}, \ldots, u_{k+b}\}$ representing all the combinations of the coding parameters from u_{k-a+1} to u_{k+b}, and the decision parameter as $x_k = u_{k-a}$. This establishes the recursion relationship for the DP algorithm, since the Markov property holds according to (3.17). With a given initial and ending states, we can recursively solve this problem.[3] This way, we reduce the original M-dimensional optimization problem in (3.15) into an $[M(a + b + 1)]$-dimensional optimization problems in (3.17). If the former is to be solved using exhaustive search, it leads to time complexity of $O(|\mathcal{U}|^M)$, while the latter has time complexity $O(M \cdot |\mathcal{U}|^{(a+b+1)})$, as the search space at stage k is $\{u_{k-a}, \ldots, u_{k+b}\}$. When $(a + b + 1)$ is much smaller than M the DP algorithm results in significant computational savings.

The parameter $(a + b)$ is referred to as the dependency order of the DP algorithm in (3.17), which clearly determines the complexity of the DP solution. For example, if the order is 0, then the optimization at each stage is carried out without considering other stages. If the order is 1, e.g., $a = 1$ and $b = 0$, the return of the kth stage (i.e., g_k) only depends on the decisions made at the previous stage and the current stage (i.e., the coding parameters for packets $k - 1$ and k). We then have

$$J_k(u_k) = \min_{u_{k-1} \in \mathcal{U}} \{g_k(u_{k-1}, u_k) + J_k(u_{k-1})\}. \qquad (3.18)$$

Thus at stage k the decision is u_{k-1} and the state is u_k. This means that when processing packet k, by evaluating all combinations of u_{k-1} and u_k, we can keep the optimal paths ending at each u_k, because the future objective functions do not depend on the decisions made on u_{k-1}. This is a first-order DP, which is essentially a shortest path algorithm, when considering the return at each stage g_k as the path length.

In the above analysis, we focus on how to apply DP to the relaxed problem (3.10). It is worth emphasizing here that the use of DP is not limited to a relaxed problem. Actually, for a given problem, as long as we can identify the stages and the associated states and decisions, we can solve the problem using DP. However, how efficient the DP algorithm will be depends

---

[3]For this particular structure, no initial and ending states are needed, since the decision parameter is one of the states at stage 0 and M.

on the size of the state space at each stage (besides the number of stages), which relates to the order of the DP problem and the decision space. If we directly solve the constrained problem as in (3.8), due to the dependency introduced by the rate constraint, the dependency of the distortion (the return in the DP algorithm) usually cannot be limited with respect to the decision parameters that lead to very large state space at each stage. For example, if the dependency cannot be limited at all, such that the selection of coding parameters for one packet depends on the coding parameters for all the other packets, the search state space will be all the possible combinations of the coding parameters for the M packets, which essentially turns the problem into an exhaustive search solution leading to very high computational complexity.

In using the Lagrangian relaxation method, we can limit the dependency of the objective function by transferring the dependency introduced by the rate constraint to the process of the Lagrange multiplier search (recall that in using Lagrangian relaxation, we break the original constrained optimization problem into two stages: one to find the appropriate Lagrange multiplier and the other to solve the resulting unconstrained minimization problem). Our goal is to solve the original constrained problem with the minimum overall computational complexity. Which method to choose depends entirely on whether we can define the stages and states for the problem, such that the search state space at each stage has limited size in order to efficiently solve the problem using a DP scheme. One example in [3] shows that in some cases, solving the original constrained problem directly using DP might be a better choice than relaxing the problem.

We illustrated above the basic idea of the DP algorithm. The recursive relationship given above is a *forward* DP solution, i.e., we solve the multistage problem starting from the initial stage. Another way to solve it is to proceed backward from the last stage, which is called *backward* DP. It has the recursion form $J_k(s_k) = \max\{r_k(s_k, x_k) + J_{k-1}(s_{k-1})\}$. Backward DP may be more widely used than forward DP. In addition, DP schemes can be classified into deterministic and stochastic DP schemes. The difference is that in the stochastic DP scheme, the return or the next state might be known with a certain probability. If the return is only known as a probability, by converting the return to an expected return based on the probability, it essentially becomes a deterministic DP scheme if we treat the expected return as the new return—that is, there will be no difference in these two approaches. However, if the transition of states is not known for certain, there will be major difference between stochastic DP and deterministic DP in that for the latter, the complete decision path is known. In a stochastic DP, the actual decision path will depend on the way the random aspects play out. Because of this, "solving" a stochastic dynamic program involves giving a decision rule for every possible state, not just along an optimal path. Again, readers can refer to [104] for a detailed discussion of DP.

CHAPTER 4

Error-Resilient Video Coding

If source coding removes all the redundancy between the source symbols and achieves entropy, a single error occurring at the source will introduce a great amount of distortion. In other words, ideal source coding is not robust to channel errors. In addition, designing an ideal or near-ideal source coder is complicated, especially for video signals, because video signals are time varying and have memory, and their statistical distribution may not be available during encoding (especially for live video applications). Thus, redundancy certainly remains after source coding. Rather than aiming at removing the source redundancy completely, we should make use of it. On the other hand, as mentioned in Chapter 3, we understand that the nature of JSCC is to optimally add redundancy at both the source- and channel-coding levels. Thus we need to regard the remaining redundancy between the source symbols after source coding as an implicit form of channel coding [94]. Actually, when JSCC is involved, source coding and channel coding sometimes can hardly be differentiated.

Generally speaking, the redundancy added should prevent error propagation, limit the distortion caused by packet losses, and facilitate error detection, recovery, and concealment at the receiver. In order to "optimally" add redundancy during source coding so as to maximize error-resilience efficiency, error-resilient source coding should adapt to the application requirements such as computational capacity, delay requirements, and channel characteristics.

Before reviewing the error-resilient source coding components, we briefly outline the video compression standards, highlight the key technologies, and introduce the necessary terminology for discussion of error-resilient source coding. At the end, we focus on the discussion of optimal mode selection, which represents the error-resilient source coding methodology that is used throughout this monograph, as an example for illustrating how to achieve optimal error resilient coding.

4.1   VIDEO COMPRESSION STANDARDS

In this section, we briefly describe one of the most widely used video coding techniques, that of hybrid block-based motion-compensated (HBMC) video coding. In the past, due to the significant developments in digital video applications, several successful standards have emerged thanks to the efforts from academia and industry. There are two main families of video compression standards: the H.26× family and the moving picture experts group (MPEG) family. These standards are application oriented and address a wide range of issues such as bit rate, complexity, picture quality, and error resilience.

The newest standard is H.264/AVC, aiming to provide the state-of-the-art compression technologies.[1] It is the result of the merger between the MPEG-4 group and the ITU H.26L committee in 2001, known as JVT (Joint Video Team), and is a logical extension of the previous standards adopted by the two groups. Thus, it is also called H.264, AVC or MPEG-4 part 10 [105]. For an overview and comparison of the video standards, see [106]. It is important to note that all standards specify the decoder only, i.e., they standardize the syntax for the representation of the encoded bitstream and define the decoding process, but leave substantial flexibility in the design of the encoder. This approach to standardization allows for maximizing the latitude in optimizing the encoder for specific applications [105].

All the above-mentioned video compression standards are based on the HBMC approach and share the same block diagram, as shown in Fig. 4.1. Each video frame is presented by block-shaped units of associated luma and chroma samples (16×16 region) called MBs (macroblocks).

As shown in Fig. 4.1(a), the core of the encoder is motion compensated prediction (MCP). The first step in MCP is motion estimation (ME), aiming to find the region in the previously reconstructed frame that best matches each MB in the current frame.[2] The offset between the MB and the prediction region is known as the motion vector. The motion vectors form the motion field, which is differentially entropy encoded. The second step in MCP is motion compensation (MC), where the reference frame is predicted by applying the motion field to the previously reconstructed frame. The prediction error, known as the displaced frame difference (DFD), is obtained by subtracting the reference frame from the current frame.

Following MCP, the DFD is processed by three major blocks, namely, transform, quantization, and entropy coding. The key reason for using a transform is to decorrelate the data so that the associated energy in the transform domain is more compactly represented and thus the resulting transform coefficients are easier to encode. The discrete cosine transform (DCT) is one of the most widely used transforms in image and video coding due to its high transform coding gain and low computational complexity. Quantization introduces loss of information, and is the primary source of the compression gain. Quantized coefficients are entropy encoded, e.g., using Huffman or arithmetic coding. The DFD is first divided into 8×8 blocks, and

---

[1] A yet newer standard for scalable video compression, scalable video coding (SVC), which is also based on H.264, is about to be finalized.

[2] The H.264/AVC standard supports more flexibility in the selection of motion compensation block sizes and shapes than any previous standards, with a minimum luma motion compensation block size as small as 4×4.

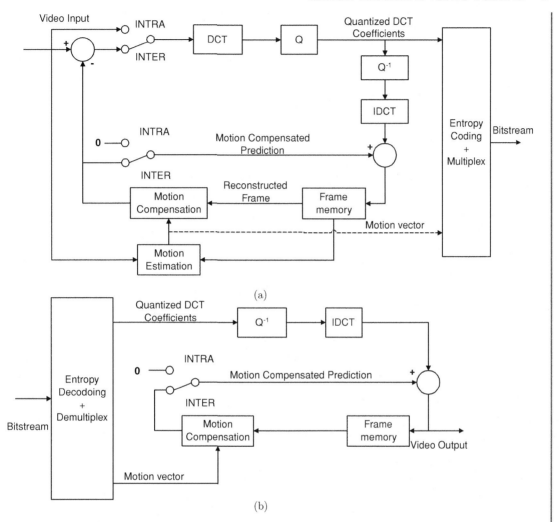

FIGURE 4.1: Hybrid block-based motion-compensated video (a) encoder and (b) decoder.

the DCT is then applied to each block, with the resulting coefficients quantized.[3] In most block-based motion-compensated (BMC) standards, a given MB can be intraframe coded, interframe coded using motion compensated prediction, or simply replicated from the previously decoded frame. These prediction modes are denoted as Intra, Inter, and Skip modes, respectively. Quantization and coding are performed differently for each MB according to its

---

[3]While all major prior video coding standards use a transform block size of 8 × 8, the H.264/AVC standard is based primarily on 4 × 4 transform. In addition, instead of DCT, a separable integer transform with similar properties as a 4 × 4 DCT is used.

mode. Thus, the coding parameters for each MB are typically represented by its prediction mode and the quantization parameter.

At the decoder, as shown in Fig. 4.1(b), the inverse DCT (IDCT) is applied to the quantized DCT coefficients to obtain a reconstructed version of the DFD; the reconstructed version of the current frame is obtained by adding the reconstructed DFD to the motion-compensated prediction of the current frame based on the previously reconstructed frame.

Besides DCT-based video compression, the wavelet representation provides a multiresolution/multiscale decomposition of a signal with localization in both time and frequency. One of the advantages of wavelet coders for both still images and videos is that they are free of blocking artifacts. In addition, they usually offer continuous data rate scalability.

During the last decade, the discrete wavelet transform (DWT) and subband decomposition have gained increased popularity in image coding due to the substantial contributions in [107, 108], JPEG2000 [109], and others. Recently, there has also been active research applying the DWT to video coding [110–115]. Among the above studies, 3D wavelet or subband video codecs have received special attention due to their inherent feature of full scalability [112, 113]. Until recently, the disadvantage of these approaches has been their poor coding efficiency caused by inefficient temporal filtering. A major breakthrough which has greatly improved the coding efficiency and led to renewed efforts toward the standardization of wavelet-based scalable video coders has come from the contributions of combining lifting techniques with 3D wavelet or subband coding [116, 117].

4.2   ERROR-RESILIENT SOURCE CODING

In this section, we first review the video source-coding techniques for supporting error resilience. After that, we briefly review the features defined in the H.263 and H.264/AVC standards that support error-resilience. As the error-resilience modes defined in MPEG-4 are mostly shared by the other two standards, we will not discuss MPEG-4 separately.

4.2.1   General error-resilience techniques

As mentioned above, error resilience is achieved by adding redundancy bits at the source-coding level, which obviously reduces the coding efficiency. The resulting question is how to optimally add redundant bits to control the tradeoff between coding efficiency and error resilience. In order to address this question, we need to identify the steps in the source-coding chain which results in corrupted bits causing significant video quality degradation.

As discussed in Chapter 2, motion compensation introduces temporal dependencies between frames, which leads to errors in one frame propagating to future frames. In addition, the use of predictive coding for the DC coefficients and motion vectors introduces spatial dependencies within a picture. Because of the use of motion compensation, an error in one part of a picture will not only affect its neighbors in the same picture, but also the subsequent frames. The solution to error propagation is to terminate the dependency chain. Techniques such as reference picture selection (RPS), intra-MB insertion, independent segment decoding, video redundancy coding (VRC), and multiple description coding are designed for this purpose. A second approach toward error resilience is to add redundancy at the entropy coding level. Examples include reversible VLCs (RVLCs), resynchronization and data partitioning techniques, which can help limit the error propagation effect to a smaller region of the bitstream once the error is detected. The third type of error-resilient source coding tools help with error recovery or concealment of the error effects, such as flexible macroblock ordering (FMO). Finally, scalable coding, although designed primarily for the purpose of transmission, along with computation and display scalability in heterogeneous environments, can provide means for error resilience by utilizing unequal error protection (UEP) through prioritized QoS transmission [13,118]. Next we provide a few more details on the techniques that provide error resilience.

- Data partitioning: This functionality is suitable for wireless channels where the bit error rate is relatively high. In traditional packetization, one MB of data including the differentially encoded motion vectors and DCT coefficients are packetized together, followed by the data of the next MB. However, in using the data partitioning mode, the data of the same type of all MBs in one packet are grouped together to form a logical unit with an additional synchronization marker inserted between different logical units. This mode enables a finer resynchronization within packets, thus providing higher level of error resiliency. That is, when an error is detected, synchronization at the decoder can be reestablished when the decoder detects the following secondary marker, thus only discarding the logical unit in which an error occurs, unlike the traditional packetization, where an error causes the decoder to discard all the data for all MBs in that packet following the detected error. One typical data partitioning syntax defined in MPEG-4 is shown in Fig. 4.2. The H.264/AVC syntax allows each slice to be separated into up to three different partitions. This functionality is also defined in MPEG-4 and H.263++ Annex V with different syntax definitions.
  In one packet, these logical units are typically of different importance. For example, packet headers typically represent the most important unit, followed by motion vectors and DCT coefficients. Data partitioning can be beneficial when error concealment

Resync Marker	MB Index	Quant Info	HEC and Header Repetition (if present)	Motion Vector	Motion Marker	Texture (DCT)	Resync Marker	...

FIGURE 4.2: Packet structure syntax for data partitioning in MPEG-4.

is used. For example, when the error-concealment strategy described in Chapter 2 is used, the received motion vectors can be used for concealing the neighboring blocks in the lost packet. In addition, if the partitioning mode is combined with UEP, e.g., by adding redundant bits or packets to provide higher protection for the more important logical units such as packet header and motion vectors, higher levels of error resiliency can be achieved. Such work can be found in [119].

- RVLC: Reversible VLCs enable decoding in both forward and backward directions, when errors are detected. In this way, more data can be salvaged since only the portion between the first MB in which an error was detected in both the forward and backward directions is discarded. This mode enhances error resiliency by sacrificing coding efficiency using a symmetric code table. RVLCs are defined in H.263++ Annex V [120], where packet headers and motion vectors can be encoded using RVLCs, but the DCT coefficients are still coded with the table used in the baseline, since the corruption of DCT information usually has less impact on the video quality compared to packet header and motion information. In MPEG-4, however, RVLCs can be applied to the DCT coefficients as well.

- Resynchronization: As the name implies, this mode is targeted at synchronizing the operations of the encoder and decoder when errors are detected in the bit stream. It is usually combined with data partitioning. There are several approaches to resynchronization defined in MPEG-4. One is the video packet approach, which is very similar in principle to the slice structured mode in H.263+. The other is the fixed interval synchronization approach, which requires video packets to start only at allowable and fixed interval locations in the bitstream.

- Scalable coding: Scalable or layered video coding produces a hierarchy of bitstreams, where the different parts of an encoded stream have unequal contributions to the overall quality. Scalable coding has inherent error-resilience benefits, especially if the layered property can be exploited in transmission, where, for example, the available bandwidth is partitioned to provide UEP for different layers with different importance. This approach is commonly referred to as *layered coding with transport pri-*

oritization [121]. In addition to the obvious benefits of scalability, layered coding with transport prioritization is one of the most popular and effective schemes for the facilitation of error resilience in a video transport system [45, 57]. Temporal, spatial, and SNR scalability are supported by those standards, i.e. in H.263+ Annex O [122], the fine granular scalability (FGS) in MPEG-4 [123], and the scalable video coding (SVC) in H.264/AVC [124]. The SVC in H.264/AVC is still an ongoing project.

- Multiple description coding (MDC): MDC refers to a form of compression where a signal is coded into a number of separate bitstreams, each of which is referred to as a description. MDC has two important characteristics. First, each description can be decoded independently to give a usable reconstruction of the original signal. Second, combining more correctly received descriptions improves the decoded signal quality. A point worth mentioning is that the descriptions are independent of each other and are typically of roughly equal importance. Thus, MDC does not require prioritized transmission. MDC will be beneficial if uncorrelated multiple paths are employed, since the use of multiple paths increases the chance of receiving at least one of them correctly. The disadvantage of MDC is its low compression efficiency compared to conventional single description coding (SDC) [125].

- Video redundancy coding (VRC): Video redundancy coding supports error resiliency by limiting the temporal dependencies between frames introduced by motion compensation. In this scenario, the video sequence is divided into two or more subsequences named *threads*. Each frame is assigned to one of the threads in a round-robin fashion and the threads are encoded independently of each other. In regular intervals, all threads converge into a so-called Sync frame, which serves as the synchronization point from which the new threads start. In this case, the picture damages in one thread will not affect the other thread(s), which can still be used to predict the next Sync frame. This technique usually outperforms I-frames insertion since Sync frames can always be generated from the intact thread, if any, without degradation.

Next we briefly describe the error-resilience features defined in H.263 and H.264/AVC. The general techniques described above will not be repeated even if they are covered by the two standards.

4.2.2 Error-resilience features in H.263+/H.263++/H.264

There are several additional features defined in H.263+, H.263++, and H.264/AVC aiming at supporting error resilience.

- Slice structure: This mode is defined in H.263+ Annex K replacing the GOB concept in baseline H.263. Each slice in a picture consists of a group of MBs, and these MBs can be arranged either in scanning order or in a rectangle shape. The reason why this mode provides error resilience can be justified in several ways. First, slices are independently decodable without using information from other slices (except for the information in the picture header), which helps limit the region affected by errors and reduce error propagation. Second, the slice header itself serves as a resynchronization marker, thus further reducing the loss probability for each MB. Third, slice sizes are highly flexible, and can be sent and received in any order relative to each other, which can help minimize latency in lossy environment. For these reasons, this mode is also defined in the H.264/AVC standard.

- Independent segment decoding: The independent segment decoding mode is defined in H.263+ Annex R. In this mode, picture segment (defined as a slice, a GOB, or a number of consecutive GOBs) boundaries are enforced by not allowing dependencies across the segment. This mode limits error propagation between well-defined spatial parts of a picture, thus enhancing the error-resiliency capabilities.

- Reference picture selection (RPS): The RPS mode is defined in H.263+ Annex N, which allows the encoder to select an earlier picture rather than the immediate previous picture as the reference in encoding the current picture. The RPS mode can also be applied to individual segments rather than full pictures. The information as to which picture is selected as the reference is conveyed in the picture/segment header. The VRC technique discussed above can be achieved through this mode.

  When a feedback channel is available, the error-resilience capability can be greatly enhanced by using this mode. For example, if the sender is informed by the receiver through a NACK that one frame is lost or corrupted during transmission, the encoder may choose not to use this picture for future prediction, and instead choose an unaffected picture as the reference. If a feedback channel is not available, this mode can still be employed to provide error-resiliency capability, e.g. through the VRC scheme discussed above.

- Flexible macroblock ordering (FMO): In the H.264/AVC standard, each *slice group* is a set of MBs defined by a *macroblock to slice group map*, which defines to which slice group each macroblock belongs. The MBs in one slice group can be in any scanning pattern, e.g., interleaved mapping, the group can contain one or more foreground and background slices and the mapping can even be a checker-board-type mapping. In addition, each slice group itself can be partitioned into one or more slices, following the raster scan order. Thus, FMO provides a very flexible tool to group MBs from different locations into one slice group, which can help deal with error-prone channels.

Note that besides the above description, the *forward error-correction* mode (Annex H)[4] is also designed for supporting error resilience [122]. As forward error correction refers to channel coding, we will discuss it in detail in the next chapter (as we can see here, the interface between source coding and channel coding becomes vague). As mentioned above, H.263+ Annex O defines temporal, spatial, and SNR scalability, and H.263++ defines data partitioning and RVLC modes in providing error resilience.

As for the H.264/AVC standard, it defines two new frame types called SP- and SI-frames, which can provide functionalities such as bitstream switching, splicing, random access, error recovery, and error resiliency [126]. This new feature can ensure drift-free switching between different representations of a video content that use different data rates. H.264/AVC also defines a new mode called *redundant pictures*, which increases error robustness by sending an additional representation of regions of pictures for which the primary representation has been lost. In addition to the above error resiliency tools defined in video coding layer (VCL), there are some other advanced features to improve error resilience defined in the network abstraction layer (NAL), such as parameter set and NAL unit syntax structures. In using the parameter set structure, the key information such as sequence header or picture header can be separately handled in a more flexible and specialized manner so as to be robust to losses. A detailed discussion can be found in [105].

4.3   OPTIMAL MODE SELECTION

As described above, there are a good number of modes defined in each video standard to provide error resilience. Different modes are designed to work under different conditions such as different applications with different bit rate requirements, or different network infrastructures with different channel error types and rates. Then one question that arises is how to optimally choose those modes in practice.

In this section, we are not going to expand the discussion on this topic, but limit our discussion on optimal mode selection that includes prediction mode (Intra, Inter, or Skip) and quantization step size for each MB or packet. Note that the Skip mode can be regarded as a special Inter mode when no error residue and motion information are coded. This topic is usually referred to as *optimal mode selection*, which has been widely studied, e.g., in [16, 17, 58, 61, 62, 127–129].

---

[4]The FEC mode is designed for Integrated Service Digital Network (ISDN) by using the BCH (511, 492) FEC code.

Mode selection algorithms have traditionally focused on RD optimized video coding for error-free environment and on single-frame BMC coding (SF-BMC) as well [16, 127, 128]. Recent research has two trends. One is a significant amount of work on mode selection using multiple-frame BMC (MF-BMC) [130]. Unlike SF-BMC, these approaches choose the reference frame from a group of previous frames. MF-BMC techniques capitalize on the correlation between multiple frames to improve compression efficiency and increase error resilience, at the cost of increased computation and larger buffers at both encoder and decoder. In this section, we focus on mode selection for SF-BMC. The other trend is to take into account the channel conditions and the error concealment used at the decoder [2–4, 6, 58, 61, 62] for the applications of video transmission over error-prone channels, which is also our main focus in this monograph.

The gist of optimal mode selection is to find the best tradeoff between coding efficiency and error robustness, since different prediction modes typically result in different levels of coding efficiency and robustness. It is worth mentioning that although our discussions are limited to the selection of the prediction mode and quantizer, the approaches and concepts described here are general to include other error resilience modes.

Our goal is to achieve the best video delivery quality with a given bit budget. This can be mathematically described as the minimization of the expected distortion D with a given bit budget R by choosing different modes, where D is calculated taking into account the channel errors. Before going in depth into the mathematical solution of this problem, it is helpful to understand how mode selection affects distortion.

Recall that the overall distortion consists of both source distortion and channel distortion, as discussed in Chapter 3. It is intuitive to see that for a given budget, the source distortion closely relates to coding efficiency, while the channel distortion closely relates to error resilience. Thus, we need to intelligently add redundant bits in the source coding to enhance the error-resilience capability with as little coding efficiency sacrifice as possible.

Generally speaking, fewer bits are usually needed to encode the DFD than its corresponding original region, since the DFD has smaller energy and entropy. For this reason, intercoding has higher compression efficiency and thus results in lower sourcecoding distortion than intracoding for the same bit budget. Intracoding, on the other hand, is more resilient to channel errors. The use of Intra coding for one MB may not contribute to the reduction of channel distortion for that MB. The contribution comes from the reduction of the effect of error propagation due to motion compensation, thus helping reduce the channel distortion of the region in the following frames that use this MB as reference. If the current MB is Inter encoded, the error that caused the corruption of its reference can affect the current MB or packet, even if this MB itself is received without error. In addition, the resulting error will propagate to the next frame if it uses this MB as reference. However, if the current MB is intraencoded, the decoder and encoder will be resynchronized if this MB is correctly received, and thus the error propagation chain will be broken. So the intramode generally results in lower channel distortion at the cost of using more bits than the intermode.

With regard to the role of the quantizer in the mode selection problem, roughly speaking, the smaller the quantization step size, the smaller the source distortion but the larger the channel distortion it may cause (for the same level of channel protection). So the quantizer also needs to be intelligently selected to achieve the best tradeoff.

Returning to the RD optimization problem for video source coding, the ultimate objective is to obtain a video of minimum distortion D given the rate constraint R. Note that the distortion D and the bit rate R here are for the whole video sequence. This is a very complicated problem. The first difficulty is due to the huge size of the set of the operational parameters for the whole video sequence considering the available modes for each MB. In tackling the mode selection problem, the focus has traditionally been on offline applications that do not consider transmission errors, such as, for example, video storage and downloading [16, 127, 128]. In this case, the optimal mode selection problem is, in theory, still a solvable problem, by employing a multiple pass scheme. However, for video transmission over error-prone channels, transmission channels are typically time varying, so the problem cannot be solved offline. Second, real-time applications, such as conversational transmission or streaming of live video, require video coding on the fly. In addition to the time varying channel, the video source is also typically time varying. This makes the mode selection a very difficult problem, since it is computationally difficult to obtain an operationally RD optimal solution and optimal solutions based on statistical models of the time-varying source and channel do not prove to be very accurate. In addition, future video frames are not available for the encoding of the current frame(s) in real-time applications, which further complicates the problem.

Facing the complexity of the problem, the research community has developed alternative approaches to finding local optimal or near-optimal solutions. The general solution is usually a greedy optimization approach that is applied to a small moving search window, e.g., to one frame or a group of frames instead of the whole video sequence. After the solution for this set of frames is obtained, the same process is repeated for the next set of frames.

So now the objective is to minimize the total distortion D for the data in the window given the rate constraint R for the window. As discussed in Chapter 3, this can be achieved in two steps. The first step is to minimize $D + \lambda R$ for a given λ, and the second step is to search for the appropriate λ^* that meets the rate constraint, $R(\lambda^*) \leq R$. The complexity of this problem depends on the size of the set of the source-coding mode parameters and the inter-MB or interpacket dependencies, which are usually determined by the source packetization scheme and the error concealment method used at the decoder.

By limiting the interpacket dependencies, by using, for example, the sliced structure mode such that each packet is independently encoded, interpacket dependencies can be managed with a reasonable level of complexity. One example of this is the work in [3], where each row of blocks is encoded as one source packet, and the encoding modes for all MBs in that packet are the same. For error concealment, the temporal concealment strategy discussed in Section 2.3.6 is used. In this case, as discussed in Section 3.5.1, the distortion of a packet depends only on the source-coding parameters for the packet above it and itself. This work has been extended in [131], whereby different prediction modes and quantization parameters for each MB within a packet are allowed, in order to achieve higher source compression efficiency. Due to the use of different prediction schemes in source encoding, both the rate and distortion of one MB depend on the prediction mode and quantization parameter chosen for the current MB and the one immediately to its left. Such intrapacket dependencies tend to lead to higher computational complexity for the optimal mode selection problem. A fast mode selection scheme based on the H.264/AVC coder has been presented in [131], which can achieve close to optimal solution with significantly reduced complexity.

The search for the appropriate λ^* usually requires several rounds of minimizations, but this approach provides a global solution (within a convex-hull approximation) to the problem (for a small search window). Note that this approach is based on the assumption that the bit budget is given for the data in the window. It is a reasonable assumption because frame-level rate control can be usually designed separately from source coding.

Another approach is to minimize $D + \lambda R$ by heuristically adjusting λ according to the bit rate target (without recursively searching for λ). Recall that as discussed in Chapter 3, the role of λ is to control the tradeoff between D and R. By minimizing $D + \lambda R$ for different values of λ we end up with different values for the rate R. Thus this approach inherently combines the frame-level rate control with the problem of mode selection for each MB, when λ for each frame is adjusted according to the bit rate budget (or the average quantization step size) for that frame, as determined by the rate controller. It can save computations by avoiding minimizing $D + \lambda R$ for several rounds with different values for λ. Examples of this type of approach can be found in [58, 61, 62].

Different models have been developed to characterize λ in terms of the bit rate R. For video compression without considering transmission errors, one such model developed in [128] is given by

$$\lambda = c \times \left(\frac{Q}{2} \right)^2, \tag{4.1}$$

where Q is the quantization step size for the block (note that the quantization parameter $QP = Q/2$ in H.263), and c is a constant. The optimal value of c is obtained experimentally by varying λ and observing the relationship between λ and the resulting average value of Q. Such value depends on the specific video encoder and the specific video coding mode being considered. In [128], $c = 0.85$ appears to be a reasonable value to characterize the functional relationship between the MB quantization stepsize and λ, while in [58], $c = 0.45$ is chosen. It has also been shown in [58] that this model works well for video transmission over error-prone channels, an assumption which has been verified for packet loss rates for up to 20%. In [61], the problem of optimal mode selection is directly related to the rate control problem via the definition of λ, which is updated per frame according to the "buffer status" as [127],

$$\lambda_{n+1} = \lambda_n \left(1 + \alpha \left(\sum_{i=1}^{n} R_i - n R_{\text{target}} \right) \right),\tag{4.2}$$

where R_{target} is the target bit rate, R_i are the bits used for the ith frame, n is the frame index, and α is given by $\alpha = \frac{1}{5 R_{\text{target}}}$. Note that in these two studies, the mode being considered only includes the prediction mode (Intra or Inter) for each MB, while the quantizer for each MB is not part of the optimal mode selection. In [58], the quantizer is explicitly chosen by the rate controller, while in [61], λ is adjusted by the target rate according to (4.2).

In this chapter, we have discussed how to optimally add redundancy at the source coding level to make the encoded bitstream robust to errors. Next we discuss how to achieve the same objective at the channel coding level.

CHAPTER 5

Channel Modeling and Channel Coding

As discussed in Chapter 3, the nature of JSCC is to optimally add redundancy at the source-coding level, which is known as error-resilient source coding, and at the channel coding level, which is known as channel coding. In Chapter 4, we have discussed the former. In this chapter, we study the latter. We first discuss channel models and then channel-coding techniques. Our focus is on the models and the techniques used for video transmission applications.

5.1   CHANNEL MODELS

The development of mathematical models that accurately capture the properties of a transmission channel is a very challenging but extremely important topic. The challenge comes from the time-varying nature of channels. The importance stems from the fact that for better video delivery performance, the end system design must be adaptive to the changing channel conditions and thus the performance of JSCC usually relies heavily on the accuracy of the channel state information estimate.

For video applications, at the application layer, the QoS is usually measured objectively by the end-to-end distortion. As discussed in Chapter 2, the end-to-end distortion is calculated according to the probability of source packet loss and delay. Thus, for video applications, two fundamental properties of the communication channel as seen at the application layer are the probability of packet loss and the delay allowed for each packet to reach the destination.

For video transmission over a network through multiple network protocol layers, the channel can be modeled at different layers. The QoS parameters at the lower layers, however, may not always directly reflect the QoS requirement by the application layer. This might not be a problem for wired network, where channel errors usually appear in the form of packet loss and truncation. For wired channels such as the Internet, the channel is modeled in a straightforward fashion at the network layer (i.e., the IP layer), since packets with errors are discarded at the link layer and are therefore not forwarded to the network layer. For wireless channels, however, besides packet loss and packet truncation, bit error is another common type of error [132].

For this reason, channels are usually modeled at the physical layer for video transmission over wireless channels. Thus for wireless networks, mechanisms that map the QoS parameters at the lower layers to those at the application layer are generally required in order to coordinate the effective adaptation of QoS parameters at the video application layer [30].

We next discuss channel models for the Internet and wireless channels, respectively.

5.1.1   Internet

In the Internet, packet loss and truncation are the typical forms of channel errors. In addition, queuing delays in the network can be a significant delay component. The Internet, therefore, can be modeled as an independent time-invariant packet erasure channel with random delays, as in [48]. In real-time video applications, a packet is typically considered lost and discarded if it does not arrive at the decoder before its intended playback time. Thus the packet loss probability is made up of two components: the packet loss probability in the network and the probability that the packet experiences excessive delay. Combining these two factors, the overall probability of loss for packet k is given by

$$\rho_k = \epsilon_k + (1 - \epsilon_k) \cdot \nu_k, \tag{5.1}$$

where ϵ_k is the probability of packet loss in the network and ν_k is the probability of packet loss due to excessive delay. We have $\nu_k = \Pr\{\Delta T_n(k) > \tau\}$, where $\Delta T_n(k)$ is the network delay for packet k, and τ is the maximum allowable network delay for this packet. This is shown in Fig. 5.1, where the probability density function (pdf) of network delay τ is plotted by taking into account packet loss.

Packet losses in the network ϵ_k can be modeled in various ways, e.g., a Bernoulli process or a two-state or a kth-order Markov chain can be used [133]. For example, Fig. 5.2 shows a

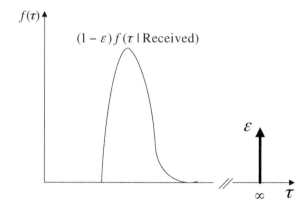

FIGURE 5.1: Probability density function of network delay taking into account packet loss.

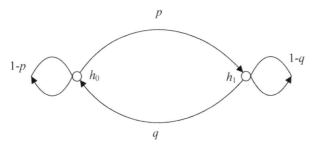

FIGURE 5.2: A two state Markov model.

two-state Markov model with channel states h_0 and h_1. The model is characterized by the channel state transition matrix defined as $A = \left| \begin{smallmatrix} 1-p & p \\ q & 1-q \end{smallmatrix} \right|$, where p and q are the probabilities of channel state transition from state h_0 to h_1 and from h_1 to h_0, respectively. The steady channel state probability is therefore computed as $\pi_0 = \frac{q}{p+q}$ and $\pi_1 = \frac{p}{p+q}$ for states h_0 and h_1, respectively.

The network delay may also be randomly varying and follow a self-similar law where the underlying distributions are heavily tailed rather than following a Poisson distribution [134, 135]. One relatively simple model used for characterizing packet delay in a network is the *shifted Gamma distribution* with rightward shift γ and parameters n and α, defined as

$$f(\tau | \text{received}) = \frac{\alpha}{\Gamma(n)} (\alpha(\tau - \gamma))^{(n-1)} e^{-\alpha(\tau - \gamma)}$$
$$\text{for} \quad \tau \geq \gamma, \tag{5.2}$$

where $\Gamma(\cdot)$ denotes the Gamma distribution. This model applies when a flow is sent through n routers, each modeled as an M/M/1 queue with a service rate α and a total end-to-end processing time of γ [48]. Short-term estimates of n, γ, and α can be obtained by periodically estimating the mean and variance of the forward trip time. For more details, see [11, 48].

5.1.2 Wireless Channel

Compared to their wire-line counterparts, wireless channels exhibit higher bit error rates, typically have a smaller bandwidth, and experience multipath fading and shadowing effects. In this subsection, we will not address the details of how to model wireless channels at the physical-layer. Our focus here is on how the physical-layer channel state information can be translated into the QoS parameters such as packet loss and delay at the link layer. How the application-layer packet loss probability (the QoS parameters needed to calculate video distortion) can be characterized by the link-layer packet loss probability depends on the packetization schemes used at the transport or link layer [6].

For IP-based wireless networks, as with the Internet, the wireless channel at the IP level can be treated as a packet erasure channel, as is "seen" by the application. In this setting, the probability of packet loss can be modeled as a function of the transmission power used in sending each packet and the channel state information (CSI). Specifically, for a fixed transmission rate, increasing the transmission power will increase the received SNR and result in a smaller probability of packet loss. This relationship could be modeled analytically or determined empirically.

As an example of the former, an analytical model based on the notion of outage capacity [136] is used in [4]. In this model, a packet is lost whenever the fading realization results in the channel having a capacity less than the transmission rate, i.e., $\rho_k = \Pr(C(H_k, P_k) \leq R)$, where C is the Shannon capacity, H_k is the random variable representing the channel's fading, P_k is the transmission power used for packet k, and R is the transmission rate (in source bits per sec). Note that here we assume that the channel fading stays fixed during one packet's transmission time. The specific expression for ρ_k depends on how the physical channel is modeled and knowledge of the channel states at the transmitter and receiver. For example, let us assume that the H_k's are independent and identically distributed (i.i.d) and that the distribution of H_k is known at the transmitter but the actual realization is not known. For a Rayleigh fading channel, H_k will have an exponential distribution and mean $E[H]$, thus the probability of packet loss ρ_k can be written as

$$\rho_k = 1 - \exp\left(\frac{1}{P_k S(\theta_k)}(2^{R/W} - 1)\right), \tag{5.3}$$

where $S(\theta_k) = \frac{N_0 W}{E[H]}$ is the normalized expected SNR given the fading level, θ_k, N_0 is noise power spectrum density and W is the channel bandwidth in Hz. Note that the derivation of the above channel model relies on the assumption of ideal channel-coding.

Another way to characterize the channel state is to use bounds for the bit error rate with respect to a given modulation and coding scheme. Consider using uncoded BPSK over a flat Rayleigh fading channel plus an additive white gaussian noise (AWGN) process. The average BER, p_e, assuming ideal interleaving, can be expressed as

$$p_e = \frac{1}{2}\left(1 - \sqrt{\frac{aE_b}{N_0 + aE_b}}\right), \tag{5.4}$$

where E_b is the bit energy, and a is the expected value of the square of the Rayleigh distributed channel gain [137]. Likewise, for an AWGN channel, the BER can be written as

$$p_e = Q\left(\sqrt{\frac{2E_b}{N_0}}\right), \tag{5.5}$$

where $Q(x) = \frac{1}{\sqrt{2\pi}} \int_x^\infty e^{\frac{-x^2}{2}} dx$ [137]. Let m be the symbol length in bits (for example, m=4 or 8) under the assumption of independent bit errors. The symbol error probability can then be written as $p_s = 1 - (1 - p_e)^m$.

Usually in wireless channels, video packets are protected by channel codes through redundant symbols within packets. A packet will be treated as lost if the corrupted symbol in this packet cannot be recovered. Assuming independent bit errors (i.e., the additive noise and fading are each i.i.d. and independent of each other), the loss probability for a transport packet in the wireless channel can be calculated as

$$\rho_k = 1 - (1 - p_b)^{B_k}, \tag{5.6}$$

where p_b is the BER after channel decoding and B_k is the source packet size in bits.

The relationship between p_b and p_e depends on the specific channel codes employed. For example, in [6, 63], rate-compatible punctured convolutional (RCPC) codes are used to perform link-layer protection. In this case, either the theoretical bounds or simulation methods for deriving BER for RCPC codes can be used, both of which can be found in [138, 139]. Note that the probability of packet loss ρ_k is a function of the transmission power level, the packet length, and the channel-coding rate selected for this packet (since p_b is a function of the channel BER p_e and channel-coding rate r_k).

In deriving the above models, we assumed independent bit errors based on perfect interleaving. Interleaving introduces complexity and delay and perfect interleaving is not achievable in a practical system, especially for real-time multimedia applications. In addition, the radio channel BER given in (5.4) and (5.5) is essentially the average BER, which is a long-term parameter. In order to account for the nature of bursty channel errors, Markov models have been widely used. For example, in the classical two-state Gilbert–Elliott model [140], channel states can be represented by one good state and one bad state with different associated BERs. This can be illustrated in Fig. 5.2, where state h_0 represents the bad state and state h_1 represents the good state. The average bursty length is $1/q$.

A more accurate model to characterize a fading channel is a finite-state Markov channel (FSMC) model [141,142]. For example, in [142], a Rayleigh flat-fading channel is modeled by an FSMC model for a packet transmission system, where each state is characterized by different bit error rates or receiver SNR. The average duration of the states is assumed to be equal to a constant, which depends on the channel fading speed. The number of states is determined by this constant, so as to ensure that each received packet completely falls into one state and the following packet only stays at the current state or one of the two neighboring states.

In the above discussion, we showed how the probability of packet loss above the link layer can be derived based on a physical-layer channel model. For real-time video transmission over a wireless network, however, besides packet loss, queueing delay is another QoS parameter that needs to be taken into account. Those physical-layer channel models do not explicitly reflect how the link-layer QoS parameters, such as the bit rate and delay, can be derived from the physical-layer channel parameters. In fact, it is very difficult to do so, because the deviations of these link-layer QoS parameters need an analysis of the queueing behavior of the connection. To achieve this, a link-layer channel model is usually required that can directly characterize the link-layer QoS parameters especially the queueing delay behavior. In the *effective capacity* (EC) model developed in [31], a wireless link is modeled by two EC functions, namely, the probability of a nonempty buffer, $\gamma = \Pr\{D(t) > 0\}$ (where $D(t)$ is the experienced delay by a source packet at time t), and the QoS exponent of a connection, θ. While γ reflects the marginal cumulative distribution function (CDF) of the underlying wireless channel, θ corresponds to the Doppler spectrum of the underlying physical-layer channel. The pair of functions $\{\gamma, \theta\}$ is the model that characterizes the link-layer channel model. These two functions can be estimated from the physical-layer channel model (such as the measurement of SNR or channel capacity) through a simple and efficient algorithm proposed in [31].

As already discussed in Section 5.1, when the packet delay is taken into account, the probability of packet loss can be generally given by (5.1). With the developed EC function, based on the theory of large deviations [31], the probability that a packet experiences excessive delay, ν, is given by

$$\nu(\theta, D_{\max}) = \Pr\left\{D(t) > D_{\max}|\theta\right\} \approx \gamma e^{-\theta D_{\max}}, \tag{5.7}$$

where $D(t)$ is the experienced packet delay at time t, $D_{\max}$ is the delay bound, and θ is the QoS exponent corresponding to the guaranteed packet delay probability.

Similarly, the probability of packet loss due to buffer overflow can be written as

$$\epsilon(\phi, B_{\max}) = \Pr\left\{B(t) > B_{\max}|\phi\right\} \approx \gamma e^{-\phi B_{\max}}, \tag{5.8}$$

where $B(t)$ is the buffer occupancy at time t, $B_{\max}$ is the maximum buffer size, and ϕ is the QoS exponent corresponding to the buffer overflow probability. Note that $\phi = \theta/\mu(\kappa)$, where $\mu(\kappa)$ is the source generation rate with the QoS exponent κ. It can also be interpreted as the effective channel capacity with the QoS exponent κ, which imposes a limit on the maximum amount of data that can be transmitted over a time-varying channel with statistical QoS guarantee in (5.8). Note that κ is the QoS exponent corresponding to the probability of packet loss required by the source generation rate. For details, please refer to [30] and [31].

The derived EC model has been successfully used in [30], where the physical-layer radio channel in a small time interval g is modeled by an L-state FSMC characterized by the $L \times L$ state transition probability matrix $p_{i,j}$, the SNR φ_i in state i of the radio channel, and the transmission bandwidth W in Hz. The achievable channel transmission rate at state i is then given by

$$r_i = W \cdot \log_2(1 + \varphi_i),$$ (5.9)

and the average link-layer transmission rate for a small time interval is

$$\bar{r} = \sum_{i=1}^{L} r_i \cdot p_i,$$ (5.10)

where p_i is the probability at state i which can be directly calculated based on $p_{i,j}$. Note that the state transition probability matrix $p_{i,j}$ can be updated at the end of each time interval g. Based on the EC model described above, for this particular physical-layer channel model, it can be translated into a closed form expression for the effective bandwidth of the source and the link-layer QoS parameters such as the effective capacity μ. For details, please refer to [30].

5.2   CHANNEL CODING

In this section, we discuss the channel-coding techniques that are widely used for images and video transmission. Two basic techniques used are FEC and ARQ. As discussed in Section 2.2, each has its own benefits with regard to error robustness and network traffic load [24,54].

5.2.1   Forward error correction

As the name indicates, FEC refers to techniques where the sender adds extra information known as check or parity information to the source information in order to make the transmission more robust to channel errors. The receiver subsequently analyzes the parity information to locate and correct errors. FEC techniques have become an important channel-coding tool used in modern communication systems. One advantage of FEC techniques is that they do not require a feedback channel. In addition, these techniques improve system performance at significantly lower cost than other techniques that aim to improve channel SNR, such as increased transmitter power or antenna gain [94].

The choice of the FEC method depends on the requirements of the system and the nature of the channel. For video communications, FEC can usually be applied across packets (at the application or transport layer) and within packets (at the link layer) [143]. In interpacket FEC, parity packets are usually generated in addition to source packets to perform cross-packet FEC, which is usually achieved by erasure codes. At the link layer, redundant bits are added within a packet to perform intrapacket prediction to reduce bit errors.

As mentioned above, the Internet can usually be modeled as a packet erasure channel. For Internet applications, many researchers have considered using erasure codes to recover packet losses [45]. With such approaches, a video stream is first partitioned into segments and each segment is packetized into a group of m packets. A block code is then applied to the m packets to generate additional l redundant packets (also called parity packets) resulting in a n-packet block, where $n = m + l$. For such a code, the receiver can recover the original m packets if a sufficient number of packets in the block are received. The most commonly studied erasure codes are the Reed–Solomon (RS) codes [144]. They have good erasure correcting properties and are widely used in practice, as for example in storage devices (VCD, DVD), mobile communications, satellite communications, digital television, and high-speed modems (ADSL) [45]. Another class of erasure codes that have recently been considered for network applications are Tornado codes, which have slightly worse erasure protecting properties, but can be encoded and decoded much more efficiently than RS codes [121].

In addition to its use as an erasure code applied across packets, RS codes can also be applied within packets to provide protection from bit errors. Besides RS codes, RCPC codes have also been widely used for providing intrapacket protection, due to their flexibility in changing the code rate and simple implementation of both encoder and decoder. In 1993, Berrou et al. [145] reported on a concatenated rate 1/2 coding scheme that used iterative decoding of achieving BER= 10^{-5} at an E_b/N_0 within about 0.5 dB away from the "pragmatic" Shannon limit. That development, known as turbo coding, has received considerable attention as a result of its ability of achieving close to Shannon's bound [145,146]. Turbo codes consist of a parallel concatenated convolutional encoder and an interleaver of the encoder inputs. The turbo decoder typically uses iterative decoding, where component decoders operate independently, each feeding its soft output information as *a priori* information to the other decoder. This type of "soft decision" decoding usually provides a gain of 2 dB compared to "hard decision" decoding. There are many basic issues with regard to the code design that require further investigation, such as selection and structure of the component convolutional codes, size and selection of the input interleaver, and improved theoretical foundations for analysis of the code weight distributions and other structural properties. In order for turbo codes to be used in video transmission applications, the above-mentioned code design criteria need to take into account the specific requirements arising from these applications. For example, wireless channels characterized by multipath fading and inter-user interference will likely alter the code design criteria, which has begun to emerge from studies of the behavior of turbo codes on AWGN channels. In addition, in order to support variable-length packets with respect to video, it is necessary to modify the interleaver, e.g., a byte-aligned variable-length turbo code based on the Jet Propulsion Laboratory (JPL) interleaver is developed for video transmissions in [147].

Next we provide some details for the RS codes and RCPC codes, as they are used for JSCC simulations in this monograph. Our focus is on demonstrating how they can be applied to video applications and their performance evaluation, i.e., the calculation of packet loss or

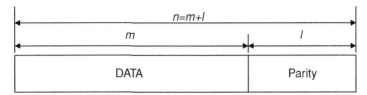

FIGURE 5.3: Illustration of an RS(n,m) codeword.

BER. For the implementation details of encoder and decoder, we refer the readers to the ample existing literature, e.g. [138, 143, 146], and the references therein.

Reed–Solomon Codes

RS codes are a subset of BCH codes and are linear block codes. Besides their good erasure correcting performance, the key benefit of RS codes is their great performance in the presence of bursty noise. An RS code is represented as RS(n, m) with s-bit symbols, where m is the number of source symbols and $l = n - m$ is the number of parity symbols. Figure 5.3 shows the graphical representation of a typical RS codeword. RS codes are based on Galois fields (GF) or finite fields. RS codes with codewords from GF(q) have length equal to $q - 1$. Given a symbol size s, the maximum codeword length for an RS code is $n = 2^s - 1$. A popular RS code is chosen from the field GF($2^8 - 1$), since each symbol can be represented as a byte. For detailed encoding and decoding operation rules and their implementations in hardware or software, refer to [139, 148] for a comprehensive tutorial.

An RS code can be used to correct both errors and erasures (an erasure occurs when the position of an error symbol is known). An RS(n, m) decoder can correct up to $(n - m)/2$ errors or up to $(n - m)$ erasures, regardless of which symbols are lost. The code rate of an RS(n, m) code is defined as m/n. The protection capability of an RS code depends on the block size n and the code rate m/n. These are limited by the extra delay introduced by FEC. The block length, n, can be determined based on the end-to-end system delay constraints [149].

The channel errors in wired links are typically in the form of packet erasures, so an RS(n, m) code applied across packets can recover up to $(n - m)$ lost packets. Thus, the block failure probability (i.e., the probability that at least one of the original m packets is in error) is

$$P_b(n, m) = 1 - \sum_{j=0}^{n-m} P(n, j) = 1 - \sum_{j=0}^{n-m} \binom{n}{j} \epsilon^j (1 - \epsilon)^{n-j}, \qquad (5.11)$$

where ϵ is the probability of packet loss before error recovery, and $P(n, j)$ represents the probability of j errors out of n transmissions.

As for wireless channels, channel coding is applied within each packet to provide protection. Source bits in a packet are first partitioned into m symbols, and then $(n - m)$ parity symbols are generated and added to the source bits to form a block. In this case, the noisy wireless channel causes symbol errors (but not erasures) within packets. As a result, the block error probability for an RS(n, m) code can be expressed as

$$P_b(n, m) = 1 - \sum_{j=0}^{(n-m)/2} P(n, j) = 1 - \sum_{j=0}^{(n-m)/2} \binom{n}{j} p_s^j (1 - p_s)^{n-j}, \qquad (5.12)$$

where p_s is the symbol error rate. The packet loss probability is then $\epsilon = P_b(n, m)$. Note that ϵ is a function of the chosen quantizer and channel-coding protection parameters for a packet, since the number of source symbols m depends on the source-coding parameters selected for this packet.

RCPC Codes

Another popular type of codes used to perform intrapacket FEC is RCPC codes [63, 143], first introduced in [138]. These codes are easy to implement, and have the property of being rate compatible, i.e., a lower rate channel code is a prefix of a higher rate channel code. A family of RCPC codes is described by the mother code of rate $1/N$ and memory M with generator tap matrix of dimension $N \times (M + 1)$. Together with N, the puncturing period G determines the range of code rates as $R = G/(G + l)$, where l can vary between 1 and $(N - 1)G$. RCPC codes are punctured codes of the mother code with puncturing matrices $\boldsymbol{a}(l) = (a_{ij}(l))$ (of dimension $N \times G$), with $a_{ij}(l) \in (0, 1)$ and 0 denoting puncturing.

The decoding of convolutional codes is most commonly achieved through the Viterbi algorithm, which is a maximum-likelihood sequence estimation algorithm. The Viterbi upper bound for the bit error probability is given by

$$p_b \leq \frac{1}{G} \sum_{d=d_{\text{free}}}^{\infty} c_d p_d,$$

where d_{free} is the free distance of the convolutional code, which is defined as the minimum Hamming distance between two distinct codewords, p_d is the probability that the wrong path at distance d is selected, and c_d is the number of paths at Hamming distance d from the all-zero path. d_{free} and c_d are parameters of the convolutional code, while p_d depends on the type of decoding (soft or hard) and the channel. Both the theoretical bounds of BER and the simulation methods to calculate BER for RCPC codes can be found in [138, 139].

5.2.2 Retransmission

Due to the end-to-end delay constraint of real-time applications, retransmissions used for video transmissions should be delay constrained. Various delay-constrained retransmission schemes

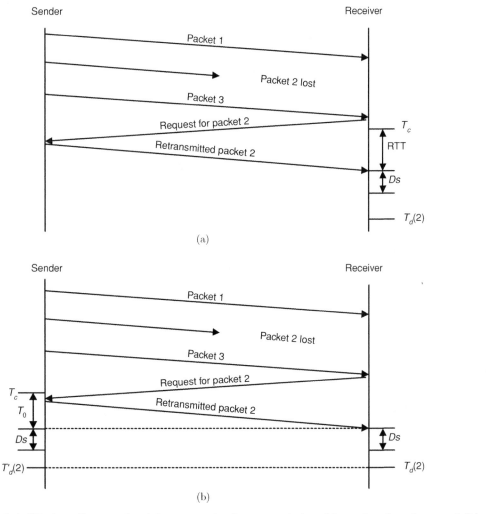

FIGURE 5.4: Timing diagram for delay-constrained retransmission (a) receiver-based control (b) sender-based control (adapted from [1]).

for unicast and multicast video are discussed in [1]. In this chapter, we focus on the unicast case, where the delay-constrained retransmissions can be classified into sender-based, receiver-based, and hybrid control, according to [1].

We illustrate the basic idea of receiver-based retransmission control in Fig. 5.4(a), where T_c is the current time, D_s is a slack term, and $T_d(n)$ is the scheduled playback time for packet n. The slack term D_s is introduced to take into account the error in estimating the RTT and other processing time, such as error correction and decoding. For a detected loss of packet n, if $T_c + \text{RTT} + D_s < T_d(n)$, which means that if the retransmitted packet n can arrive at the receiver before its playback time, the receiver sends a retransmission request of packet n to the sender. This is the case depicted in Fig. 5.4(a) for packet 2.

Different from the receiver-based control, in sender-based control, decisions are made at the sender end. The basic idea is illustrated in Fig. 5.4(b), where T_0 is the estimated forward-trip-time, and $T'_d(n)$ is an estimate of $T_d(n)$. If $T_c + T_0 + D_s < T'_d(n)$ holds, it can be expected that the retransmitted packet n will arrive at the receiver in time for playback. The hybrid control is a direct combination of the receiver-based and sender-based control, so that better performance can be achieved at the cost of higher complexity. After laying out the basic concept of delay-constraint retransmission, we next discuss how retransmission techniques are implemented in a network.

On the other hand, delay-constrained retransmission can be implemented in multiple network layers. It is well known that TCP is a reliable end-to-end transport protocol that provides reliability by means of a window-based positive acknowledgement (ACK) with a go-back-N retransmission scheme [35]. In an IP-based wireless network for the emerging 3G and 4G systems, such as CDMA2000, both radio link control (RLC) frame retransmissions and MAC frame retransmissions are allowed [150]. The current WLAN standard IEEE 802.11 also allows MAC frame retransmission [27]. Due to the strict end-to-end delay constraint, TCP is usually not preferred for real-time video communications. However, compared to transport-layer retransmission TCP, link-layer and MAC-layer retransmission techniques introduce smaller delays, because the lower layers react to the network faster than the upper layers [151]. Since delay-constrained retransmission at the link and MAC layers introduce much shorter delays, they are widely used in real-time video communications [27]. For example, researchers have been studying how many retransmissions in the MAC layer are appropriate for multimedia transmission applications in order to achieve the best tradeoff between error correction and delay [26, 27, 151].

CHAPTER 6

Internet Video Transmission

Video communications over the Internet is an active and extensively studied field of research. In this chapter, we study the JSCC problem for video transmission over the Internet. In particular, we focus on the techniques that efficiently adapt both source and network parameters to achieve the best video delivery quality with given resource constraints. In the following sections, we first address the general problem formulation presented in Chapter 3, and then go on to show how the framework is applied to different applications and network infrastructures through three case studies.

6.1 GENERAL PROBLEM FORMULATION

As mentioned in Chapter 1, a direct approach dealing with the lack of QoS in the best-effort design Internet is to use error control, where different error-control components can be implemented in different network layers. Error-control techniques for Internet video transmission in general include error-resilient source coding at the encoder, channel coding (FEC and ARQ) at the application layer, and error concealment at the receiver. In this chapter, we jointly consider a combination of these error control approaches. For error-resilient source coding, we will consider mode selection where the mode for each MB or packet consists of the prediction mode (intra, inter, or skip) and the quantizer step size. For error concealment at the decoder, we consider the strategy discussed in Chapter 2. Thus, our focus is on channel coding, i.e., for different applications and different network infrastructures, we will study which channel-coding technique is preferable and how to jointly optimize channel-coding and source-coding mode selection. The study is carried out in the resource-distortion optimization framework presented in Chapter 3.

Recall that the goal of the general formulation of resource-distortion optimization discussed in Chapter 3 is to minimize the end-to-end distortion while using a limited amount of resources and delay. This is expressed as

$$
\begin{aligned}
\min_{s \in \mathcal{S}^M, c \in \mathcal{C}^M} \quad & E[D(s, c)] \\
\text{s.t.} \quad & C(s, c) \leq C_0 \\
& T(s, c) \leq T_0,
\end{aligned}
\tag{6.1}
$$

where M is the number of packets considered in the optimization, $\mathcal{S}$ and $\mathcal{C}$ are, respectively, the set of source- and channel-coding parameters, $s = \{s_1, ..., s_M\} \in \mathcal{S}^M$ and $c = \{c_1, ..., c_M\} \in \mathcal{C}^M$ denote, respectively, the vector of source- and channel-coding parameters for the M packets, C_0 is the maximum allowable resource consumption, and T_0 is the end-to-end delay constraint. We will refer to this approach as the *minimum distortion (MD) approach* because it provides the best possible quality given cost and delay constraints. A dual formulation to (6.1) is to minimize the cost required to transmit a video frame (or group of frames) with an acceptable level of distortion and with tolerable delay, that is,

$$\min_{s \in \mathcal{S}^M, c \in \mathcal{C}^M} C(s, c)$$
$$\text{s.t.} \quad D(s, c) \leq D_0 \tag{6.2}$$
$$T(s, c) \leq T_0,$$

where D_0 is the maximum allowable distortion. We will refer to (6.2) as the *minimum cost (MC) approach*. This dual formulation is useful for applications with stringent QoS requirements. For example, (6.2) can be used to maintain a constant level of distortion by varying the resource consumption per frame [3,4].

Approaches, such as (6.1) and (6.2), in which source and channel parameters are jointly adapted, have only recently been considered. A divide and conquer approach was typically used before where these parameters were adapted separately. There are several reasons for this, one of which is the fact that the computational complexity of optimally solving (6.1) and (6.2) has been considered infeasible until recently. Advances in low-cost and high-speed computing are enabling researchers to reconsider these more complex problems. A second reason is that the solution to (6.1) and (6.2) is not straightforward, and researchers in the past have resorted to ad hoc solutions and tweaking of parameters. Recent advances in optimal joint source-network resource allocation required "innovative" application of established optimization techniques.

The formulations in (6.1) and (6.2) are meant to draw attention to the high-level similarities between various resource allocation problems. In different applications, source-coding adaptations, channel-coding adaptations, and resources have different forms. It is important to note that applying these formulations to different scenarios requires addressing the intricacies and challenges of each application. The optimization tools which are often employed to solve (6.1) and (6.2) have been discussed in Chapter 3.

In the rest of this chapter, we discuss three special cases of the general formulation. First we consider joint source coding and FEC in Section 6.2. In addition to FEC, retransmission-based error control may also be used in the form of ARQ protocols. Such protocols are only useful if the application can tolerate sufficient delay, such as in the case of on-demand streaming. When ARQ protocols are used, the decision whether to retransmit a packet or send a new one forms another channel coding parameter, which also affects the probability of packet loss as well as the transmission delay. In Section 6.3, we study hybrid application-layer error control, especially the performance of pure FEC, pure retransmission, and hybrid FEC and retransmission in achieving error protection. Finally, in Section 6.4, we consider video transmission over DiffServ networks, where we jointly consider source coding and packet classification.

6.2 JOINT SOURCE CODING AND FORWARD ERROR CORRECTION

Various JSCC approaches for video or image transmission over the Internet have been widely studied (see, for example, [1, 2, 11, 13, 58, 60, 99, 149]). As discussed in Chapter 5, FEC and ARQ are two basic error correction techniques. Of the two, FEC has been widely used for real-time video applications due to the strict delay requirement and semireliable nature of video streams [1, 13]. When considering joint source coding and interpacket FEC, the JSCC problem becomes a special case of the general resource-distortion framework in (6.1), and can be written as

$$\min_{s \in \mathcal{S}^M, c \in \mathcal{C}^M} E[D(s, c)]$$
$$\text{s.t.} \quad T(s, c) \le T_0. \tag{6.3}$$

As mentioned in Chapter 2, the appropriate way of calculating the loss probability per packet depends on the chosen FEC as well as the way transport packets are formed from the available video packets. Next, we give one example where the source packet is a video slice (a group of blocks).

Figure 6.1 illustrates a packetization scheme for a frame, where one row corresponds to one packet. In this packetization scheme, one video slice is directly packetized into one transport packet by the attachment of a transport packet header. Since the source packet sizes (shown by the shaded area in Fig. 6.1) are usually different, the maximum packet size of a block (a group of packets protected by one RS code) is determined first, and then all packets are padded with stuffing bits in the tail part to make their sizes equal. The stuffing bits are removed after the parity codes are generated and thus are not transmitted. The resulting parity packets are all of the same size (the maximum packet size mentioned above). Each source packet in Fig. 6.1 is protected by an RS(N, M) code, where M is the number of video packets and N is the number of total packets including parity packets.

In this case, the channel-coding parameter c represents the choice of the RS code rate and the source-coding parameter s includes the prediction mode and quantizer for each video packet. The optimal JSCC solution, i.e., the optimal bit allocation as well as the optimal

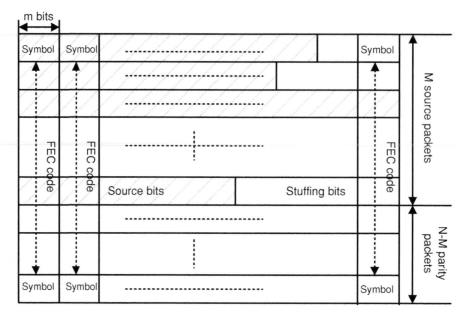

FIGURE 6.1: Illustration of a packetization scheme for inter packet FEC.

error-resilient source coding and FEC, can be obtained by solving (6.3) using the techniques introduced in Chapter 3.

To illustrate the advantage of the JSCC approach, we compare two systems [2]: (i) system 1, which uses the proposed framework to jointly consider error-resilient source coding and channel coding; and (ii) system 2, which performs error-resilient source coding, but with fixed rate channel coding. Note that system 2 is also optimized, i.e., it performs optimal error-resilient source coding to adapt to the channel errors (with fixed rate channel coding).

In Fig. 6.2, the performance of the two systems is compared, using the QCIF Foreman test sequence coded by an H.263+ codec at a transmission rate of 480 kbps and frame rate equal to 15 fps. Here, we plot the average PSNR in dB versus different packet loss rates. It can be seen in Fig. 6.2 that system 1 outperforms system 2 at different preselected channel coding rates. In addition, system 1 is always above the envelope of the four performance curves of system 2. This is due to the flexibility of system 1, which is capable of adjusting the channel-coding rate in response to the CSI as well as the varying video content.

In Fig. 6.3, the average PSNR is plotted against the transmission rate for the two systems. It can be seen that system 1 outperforms system 2 with different preselected channel-coding rates at various transmission rates. In addition, this figure illustrates the effect of bit budget on the JSCC problem. It can be clearly seen that as the transmission rate increases (i.e., the bit budget per frame increases), the gap between the performance of system 2 (without channel

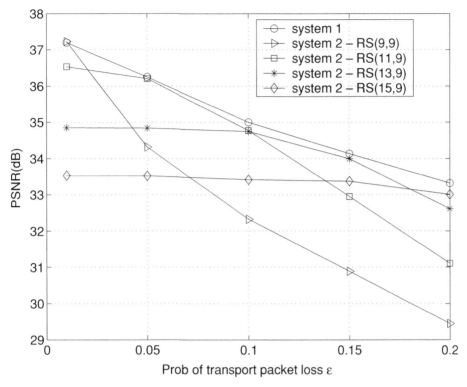

FIGURE 6.2: Average PSNR versus transport packet loss probability ϵ (QCIF Foreman sequence, transmission rate 480 kbps, coded at 15 fps) (adapted from [2]).

coding) and the other systems (with channel coding) also increases. With a low bit budget, the ability of using channel coding is restricted, because a majority of the bits is needed for source coding. When the bit budget increases, it becomes more flexible for the sender to allocate bits to the channel. As a result, the overall system performance is improved.

6.3   JOINT SOURCE CODING AND HYBRID FEC/ARQ

We now consider joint source coding and hybrid FEC and ARQ in performing optimal error control. That is, for channel coding, we consider the combination of FEC and application-layer selective retransmission, where the channel-coding parameter c in (6.1) includes the FEC rate chosen to protect each packet and the retransmission policy for each lost packet.

Hybrid FEC/retransmission has been considered in [48], where a general cost-distortion framework is proposed to study several scenarios such as DiffServ, sender-driven retransmission, and receiver-driven retransmission. A receiver-driven hybrid FEC/pseudo ARQ mechanism is proposed for Internet multimedia multicast in [24]. Such a system looks

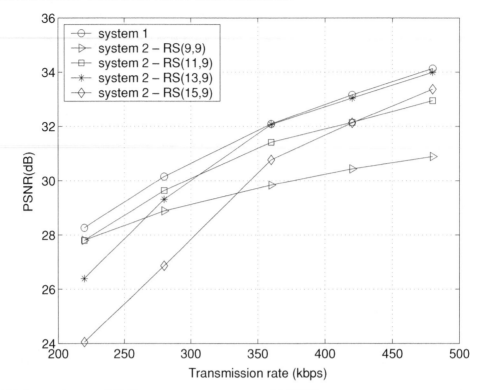

FIGURE 6.3: Average PSNR versus transmission rate (QCIF Foreman sequence, $\epsilon = 0.15$, coded at 15 fps) (adapted from [2]).

like an ordinary ARQ to the receiver and an ordinary multicast to the sender. This can be characterized as JSCC with receiver feedback. More specifically, the optimal JSCC is obtained by solving (6.3) at the receiver side, where the source-coding parameter is the number of source-coding layers (i.e., the source layers using a scalable video codec), and the channel-coding parameter is the number of channel-coding layers (i.e., parity layers). For wireless IP networks, a link-layer hybrid FEC/ARQ scheme is considered in [56] and an application-layer hybrid FEC/ARQ technique based on heuristics is presented for video transmission in [54].

In [2], optimal error control is performed by jointly considering source coding with hybrid FEC and sender-driven application-layer selective retransmission. This study is carried out with the use of (6.1), with a sliding window scheme in which lost packets are selectively retransmitted according to a rate-distortion optimized policy. In that work, the benefits from both the proposed hybrid FEC and selective retransmission are utilized since the type of error control is adapted based on the channel characteristics. We next describe such an approach in detail.

6.3.1   Problem Formulation

Assume that there are up to A frames in the sender's buffer that are eligible for retransmission (determined by the encoder's buffer size and/or the delay requirement). Following the notations in [2], let $\mathcal{Q}$ be the set of source-coding parameters, which include the prediction mode and quantization step size. The FEC parameter set is defined as $\mathcal{R} = \{(N_1, M), ..., (N_q, M)\}$, where q is the number of available code options (see Fig. 6.1). Let $\boldsymbol{\mu}$ and $\boldsymbol{\nu}$ denote the vectors of source-coding parameters and FEC parameters for one frame, respectively. The retransmission parameter for the kth source packet in frame n is denoted as $\sigma_k^{(n)} \in \{0, 1\}$, where 0 denotes no retransmission and 1 denotes retransmission. Let $\boldsymbol{\sigma}^{(n)} = \{\sigma_1^{(n)}, ..., \sigma_M^{(n)}\}$ denote the retransmission parameter vector for frame n, and $\boldsymbol{\sigma} = \{\boldsymbol{\sigma}^{(n-A)}, ..., \boldsymbol{\sigma}^{(n-1)}\}$ the vector for the past A frames. For video transmission applications, usually a higher level rate controller is used to constrain the bits, or equivalently the transmission delay for each frame. For simplicity, let $T_0^{(n)}$ be the transmission delay for the nth frame obtained from the rate controller. Following the structure of the JSCC framework in (6.1), we encompass it into the following formulation:

$$\min_{\boldsymbol{\mu} \in \mathcal{Q}, \boldsymbol{\nu} \in \mathcal{R}, \boldsymbol{\sigma} \in \mathcal{P}} \sum_{i=0}^{A} E[D^{(n-i)}] = \min_{\boldsymbol{\mu} \in \mathcal{Q}, \boldsymbol{\nu} \in \mathcal{R}, \boldsymbol{\sigma} \in \mathcal{P}} \left\{ \sum_{i=1}^{A} E[D^{(n-i)}(\boldsymbol{\sigma}^{(n-i)})] + \sum_{k=1}^{M} E[D_k^{(n)}(\boldsymbol{\mu}, \boldsymbol{\nu})] \right\}$$

$$\text{s.t.} \sum_{i=1}^{A} \sum_{k=1}^{M} \sigma_k^{(n-i)} T_k^{(n-i)} + \sum_{k=1}^{M} T_k^{(n)} \leq T_0^{(n)}. \tag{6.4}$$

The above formulation is for an optimization scheme with a sliding window of size $A+1$ frames. The optimization window shifts at the frame level instead of at the packet level, since the latter usually leads to much higher computational complexity. In addition, the packets in one frame typically have the same deadline for playback. In this formulation, when processing each frame, we assume that all the raw data for the frame are available in a buffer, and the optimization (retransmission policy for the first A frames based on feedback, and source coding and FEC for the current frame) is performed on the $A+1$ frames in the window. After the optimization is performed, the retransmitted packets and the transport packets in the current frame (including source packets and parity packets) are transmitted over the network. After the transmission of these packets, the window shifts forward by one frame, and the optimization is performed again based on the updated feedback.

When each frame is encoded, the probability of packet loss for all the past A frames is updated based on the received feedback. For example, if one packet is known to be received, its probability of loss becomes 0; if one is lost, its loss probability becomes 1 if no further retransmission for this packet has been initiated. Based on the updated probabilities of packet loss, the expected distortion of all packets in the encoder buffer is recursively recalculated as in (2.3). Since each time we do not consider reencoding of the past A frames, the complexity in updating the expected distortion is not significant.

6.3.2 Calculation of Probability of Packet Loss

In order to find the expected distortion in (2.3), the probability of packet loss ρ_k needs to be determined. The calculations of loss probability of packets in the current and past frames are different. For a packet in the current frame, the probability of packet loss can be defined as $\rho_k^{(n)} = \rho_{k,\text{FEC}}^{(n)} \rho_{k,\text{RET}}^{(n)}$, where $\rho_{k,\text{FEC}}^{(n)}$ and $\rho_{k,\text{RET}}^{(n)}$ denote, respectively, the probability of packet loss due to FEC and retransmission.

When protected by an RS(N, M) code, a source packet is regarded as lost after error recovery at the receiver only when the corresponding transport packet is lost and the block containing the lost transport packet cannot be recovered (i.e., more than $N - 1 - M$ other packets are also lost.). Therefore, $\rho_{k,\text{FEC}}^{(n)}$ is given by

$$\rho_{k,\text{FEC}}^{(n)} = \epsilon(1 - \sum_{i=0}^{N-1-M} \binom{N-1}{i} \epsilon^i (1-\epsilon)^{N-1-i}) = \sum_{i=N-M+1}^{N} \frac{i}{N} \binom{N}{i} \epsilon^i (1-\epsilon)^{N-i}, \quad (6.5)$$

where ϵ is the probability of transport packet loss. The probability of loss in future retransmissions can only be estimated since the acknowledgement information and retransmission decisions (note that lost packets are selectively retransmitted) are not available in the encoding of the current frame. As shown in [2], it can be estimated as $\rho_{k,\text{RET}}^{(n)} = \epsilon^{\tilde{m}}$, where $\tilde{m}$ denotes the estimate of the total number of retransmissions for the kth packet, m. Based on the simulation results given in [2], we have $\tilde{m} = \frac{A}{(1+\text{RTT})^2}$, where RTT is the round-trip-time in the units of one frame's duration T_F. This appears to provide good performance and is used subsequently. Note that the maximum number of available retransmission opportunities is $\lfloor A/(1 + \text{RTT}) \rfloor$.

In considering the possible retransmission of packets in the current frame, the expected additional transmission delay used for retransmission in the future should be taken into account; it is calculated by $E[\Delta T_k^{(n)}] = \sum_{k=1}^{M} \tilde{m} \rho_{k,\text{FEC}}^{(n)} T_k^{(n)}$. The delay constraint in (6.4) is modified accordingly.

For a lost packet in the past frames, we let $\rho_k^{(n-i)} = \rho_{k,\text{UPD}}^{(n-i)} \rho_{k,\text{RET}}^{(n-i)}$ for $i = 1, ..., A$, where $\rho_{k,\text{UPD}}^{(n-i)}$ is the updated probability of packet loss based on feedback and $\rho_{k,\text{RET}}^{(n-i)}$ is the probability of packet loss due to retransmissions. Assume that one past frame is protected by an RS(N, M), and L packets are lost. Let $J = L + M - N$ and V be the number of retransmitted packets in that frame. Taking into account the RS codes, the calculation of $\rho_{k,\text{RET}}^{(n-i)}$ is different for the lost packets that are either retransmitted or not. If $V < J$, we have

$$\rho_{k,\text{RET}}^{(n-i)} = \epsilon^{\sigma_k^{(n-i)}};$$

if $V = J$, we have

$$\rho_{k,\text{RET}}^{(n-i)} = \begin{cases} \epsilon & \text{if } \sigma_k^{(n-i)} = 1 \\ 1 - (1-\epsilon)^J & \text{if } \sigma_k^{(n-i)} = 0; \end{cases}$$

and if $V > J$ we have

$$\rho_{k,\text{RET}}^{(n-i)} = \begin{cases} \sum_{j=V-J+1}^{V} \frac{j}{V}\binom{V}{j}\epsilon^j(1-\epsilon)^{V-j} & \text{if } \sigma_k^{(n-i)} = 1 \\ \sum_{j=V-J+1}^{V} \binom{V}{j}\epsilon^j(1-\epsilon)^{V-j} & \text{if } \sigma_k^{(n-i)} = 0. \end{cases}$$

Note that when the above formulas are derived, in order to maintain reasonable complexity, we make a conservative assumption regarding future retransmissions of packets for which we have received negative acknowledgements. In particular, when calculating $\rho_{k,\text{RET}}^{(n-i)}$, we only consider the possibility of retransmission at the current time instant. Without this assumption, the complexity required to estimate the probability of loss based on future retransmissions of previously lost packets increases significantly and is therefore less practical.

6.3.3   Solution Algorithm

We next present our solution to this formulation. By using a Lagrange multiplier $\lambda \geq 0$, (6.4) can be converted into an unconstrained problem as

$$\min_{\mu \in \mathcal{Q}, \nu \in \mathcal{R}, \sigma \in \mathcal{P}} \sum_{i=0}^{A} J^{(n-i)}$$

$$= \min_{\mu \in \mathcal{Q}, \nu \in \mathcal{R}, \sigma \in \mathcal{P}} \left\{ \sum_{i=1}^{A} E[D_k^{(n-i)}(\sigma^{(n-i)})] + \sum_{k=1}^{M} E[D_k^{(n)}(\mu, \nu)] \right. \tag{6.6}$$

$$\left. + \lambda \left(\sum_{i=1}^{A}\sum_{k=1}^{M} \sigma_k^{(n-i)} T_k^{(n-i)} + \sum_{k=1}^{M} T_k^{(n)} \right) \right\},$$

where $J^{(n-i)} = E[D_k^{(n-i)}] + \lambda \sum_{k=1}^{M} \sigma_k^{(n-i)} T_k^{(n-i)}$. As discussed above, if we do not consider reencoding of the past A frames (i.e., we only consider retransmission of the packets in the past

A frames), the recursion of this problem is then given as follows:

$$\min_{\mu \in \mathcal{Q}, \nu \in \mathcal{R}, \sigma \in \mathcal{P}} \sum_{i=0}^{A} J^{(n-i)}$$

$$= \min_{\sigma \in \mathcal{P}} \sum_{i=1}^{A} J^{(n-i)}(\sigma^{(n-i)}) + \min_{\nu \in \mathcal{R}} \left\{ \min_{\mu \in \mathcal{Q}} \sum_{k=1}^{M} J_k^{(n)}(\mu, \nu) \right\}, \tag{6.7}$$

where $J_k^{(n)} = E[D_k^{(n)}(\mu, \nu)] + \lambda T_k^{(n)}$. There are three minimizations in (6.7). They correspond to the bit allocation for retransmission, bit allocation for FEC, and the optimal mode selection for the current frame based on the remaining delay. The minimizations with respect to σ and ν can be solved using exhaustive search, while the optimal mode selection can be found using a DP approach. Note that by using the error concealment strategy described in Section 2.3.6, the time complexity is $O(|2^L \times |\mathcal{R}| \times M \times |\mathcal{Q}|^2)$, where L is the number of lost packets in the optimization window. If the error concealment strategy does not introduce dependencies across source packets, the time complexity would be $O(|2^L \times |\mathcal{R}| \times M \times |\mathcal{Q}|)$ [17].

6.3.4   Experimental Results

In this subsection, we study the efficiency and effectiveness of these two error correction techniques, FEC and ARQ, through the JSCC framework in different network situations (such as packet loss probability and network round trip time) and application requirements (such as end-to-end delay). Four schemes are compared: (i) neither FEC nor retransmission (NFNR), (ii) pure retransmission, (iii) pure FEC, and (iv) Hybrid FEC and selective retransmission (HFSR). All four systems are optimized using the JSCC framework. Although the JSCC framework in (6.4) is general, in the simulations, we restrict a packet's retransmission only when its NAK has been received. In all experiments in this subsection, we utilize the QCIF Foreman sequence and set $A = 4$.

Sensitivity to RTT

Figure 6.4 shows the performance of the four systems in terms of PSNR versus RTT, with different probabilities of transport packet loss, ϵ. We set $R_T = 480$ kbps and $F = 15$ fps. As expected, the HFSR system offers the best overall performance. It can also be seen that the pure retransmission approach is much more sensitive to variations in the RTT than FEC. In addition, at low ϵ and low RTT, the pure retransmission approach outperforms the pure FEC system, as shown in Fig. 6.4(a). However, when the channel becomes worse and the RTT becomes larger, the pure FEC system starts to outperform the pure retransmission system, as shown in Fig. 6.4(b). This means that retransmission is suitable for those applications where the RTT is short and the channel loss rate is low, which confirms the observation in [54]. The disadvantage of retransmission when the RTT gets longer stems from two factors: (1) Given

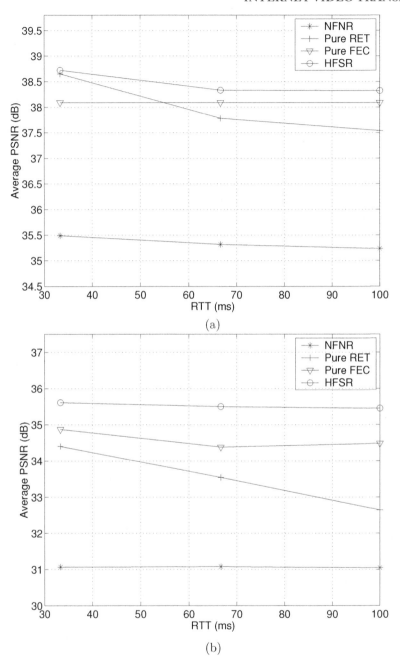

FIGURE 6.4: Average PSNR versus RTT, $R_T = 480$ kbps, $F = 15$ fps (a) ϵ =0.02 (b) $\epsilon = 0.2$ (adapted from [2]).

the same value of A, which is decided by the initial setup time T_{max}, the number of retransmission opportunities becomes smaller; (2) Errors accumulated due to error propagation from motion compensation become larger, and consequently retransmission of lost packets becomes less efficient.

Sensitivity to packet loss rate

In Fig. 6.5, we plot the performance of the four systems in terms of PSNR versus probability of transport packet loss ϵ for different values of RTT, for $R_T = 480$ kbps and $F = 15$ fps. The RTT is set equal to T_F and $3T_F$ in Fig. 6.5(a) and 6.5(b), respectively. It can be seen that the HFSR system achieves the best overall performance of the four systems under comparison. The resulting PSNR in the pure retransmission system drops faster than the pure FEC system, which implies that retransmission is more sensitive to packet loss rate. In addition, the pure retransmission system only outperforms the pure FEC system at low probability of packet loss, ϵ. When the channel loss rate is high, FEC is more efficient, since retransmission techniques require frequent retransmissions to recover from packet loss, which result in high bandwidth consumption and are also limited by the delay constraint. For example, when RTT= $3T_F$ and $A = 4$, each lost packet has only one chance for retransmission, which is not enough to recover many losses when $\epsilon = 0.3$. However, when the channel loss rate is small and the RTT is small, retransmission becomes more efficient, since FEC typically requires a fixed amount of bandwidth overhead. Consequently, in this case the pure retransmission system performs close to the HFSR system, as shown in Fig. 6.5(a).

Sensitivity to transmission rate

Figure 6.6 shows the performance of the four systems in terms of PSNR versus transmission rate when $\epsilon = 0.2$ and $F = 15$ fps. The RTT is set equal to T_F and $3T_F$ in Figs. 6.6(a) and 6.6(b), respectively. As shown in Fig. 6.6(a), when RTT= T_F, the pure retransmission system outperforms the pure FEC system by up to 0.4 dB when the transmission rate is less than 450 kbps. When the transmission rate is greater than 450 kbps, the pure FEC system outperforms the pure retransmission system by up to 0.5 dB. When the RTT becomes longer, as shown in Fig. 6.6(b), although the pure FEC system always outperforms the pure retransmission system, the difference between the two systems increases from 1.2 dB to 1.8 dB when the transmission rate increases from 240 kbps to 540 kbps, which means that FEC is more sensitive to variations in the transmission rate. These observations imply that FEC is more efficient than retransmission when the transmission rate becomes larger (resulting in a higher bit budget per frame). As discussed in Chapter 5, due to the constant overhead introduced by FEC, its use is restricted at small transmission rates. When the bit budget increases, however, the system becomes more flexible in its ability of allocating bits to the channel in order to improve the overall performance.

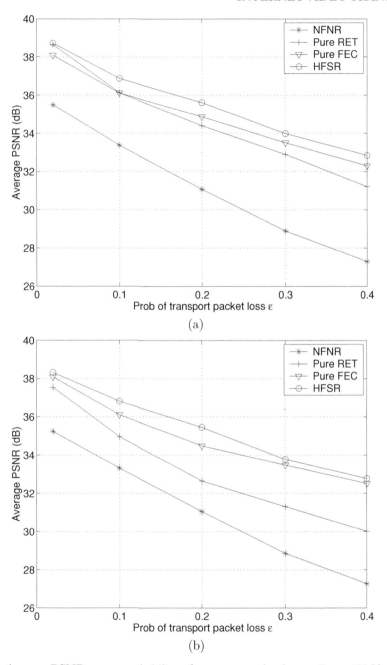

FIGURE 6.5: Average PSNR versus probability of transport packet loss ϵ, $R_T = 480$ kbps, $F = 15$ fps (a) RTT=T_F (b) RTT=$3\,T_F$ (adapted from [2]).

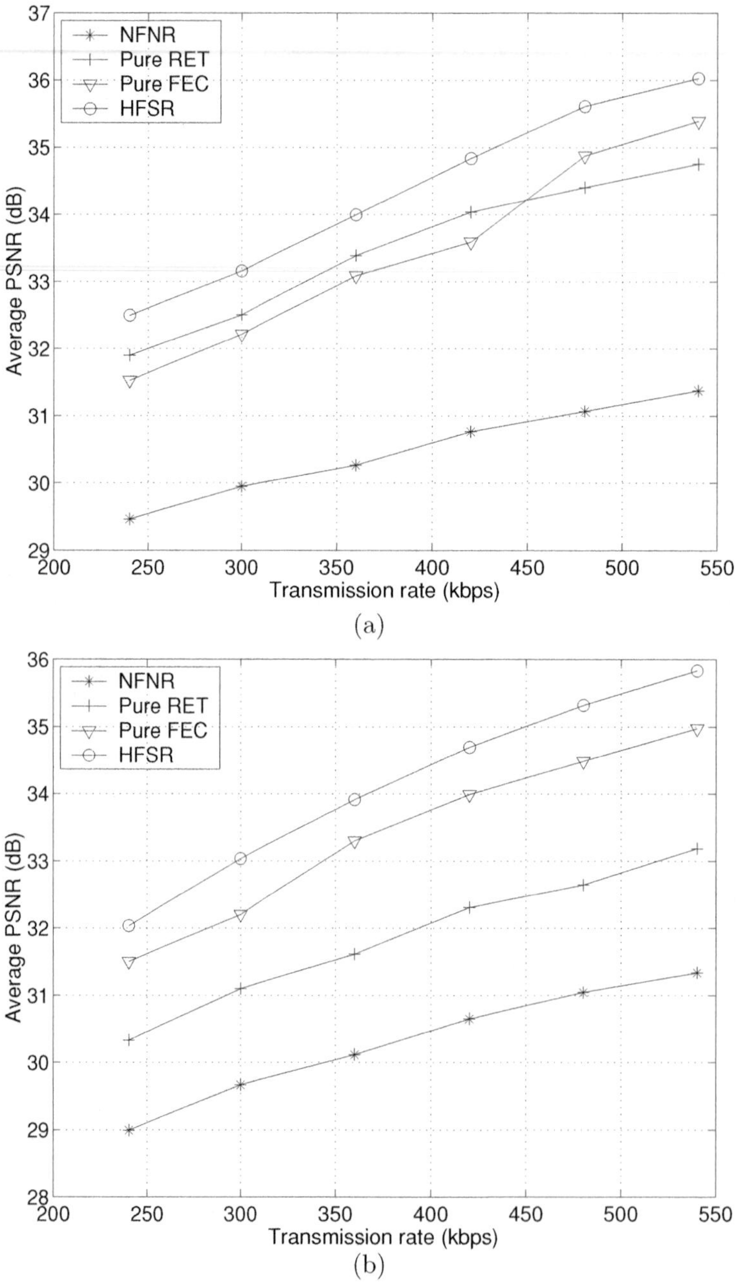

FIGURE 6.6: Average PSNR versus channel transmission rate R_T, $\epsilon = 0.2$, $F = 15$ fps (a) RTT=T_F (b) RTT=$3\,T_F$ (adapted from [2]).

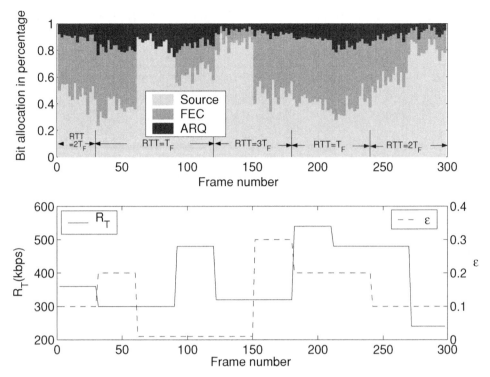

FIGURE 6.7: Average bit allocation of the HFSR system over a time-varying channel at $F = 15$ fps (adapted from [2]).

As expected again, the HFSR system achieves the best overall performance of the four systems under comparison.

Time-varying channel

In Fig. 6.7, we show how the proposed HFSR system responds to network fluctuations. The top figure shows the average bit allocation between source coding, FEC, and ARQ. The network fluctuations including variations in the channel transmission rate and packet loss probability are illustrated in the bottom graph. The variations in RTT are also indicated in the top graph. In the simulations, the sender carries out the optimization based on the currently estimated CSI. The average bit allocation is obtained based on twenty different channel realizations with the same channel characteristics. It can be seen that the system intelligently allocates bits to source coding, FEC, and ARQ, in response to the changing network conditions. For example, during the time from frame 61 to 75, when the RTT is short, the transmission rate is low, and the packet loss probability is also low, most of the bits are allocated to source coding and the remaining bits are used for ARQ. When the transmission rate increases, more bits are allocated to FEC after the 75th frame, because of the higher flexibility of the proposed system in bit allocation. However, when the packet loss probability increases after the 150th frame, more bits are needed to combat channel errors and therefore the amount of bits allocated to source coding must decrease. The observations from Fig. 6.7 further confirm what we have seen in Figs. 6.4–6.6. Thus, the proposed system performs very well in response to the network fluctuations.

Although we only showed simulation results based on the QCIF Foreman sequence, extensive experiments have been carried out and similar results were obtained using other test sequences such as Akiyo, Container, and Carphone. More experimental results can be found in [2].

In summary, retransmission is suitable for short network RTT, low probability of packet loss, and low transmission rate, while FEC is more suitable otherwise. In general, through solving (6.4), the presented hybrid FEC and selective retransmission scheme is able to identify the best combination of the two.

6.4   JOINT SOURCE CODING AND PACKET CLASSIFICATION

In this section, we study the problem of joint source-channel coding for video transmission over Diffserv (DS) networks, i.e., joint source coding and packet classification. In this case, the "channel coding" role is provided by the choice of the DS class.

Today's Internet is a best effort network, and thus does not provide any guaranteed QoS. As discussed in Chapter 2, DiffServ has recently been considered as a means of providing discriminate quality of service based on service classes. In this architecture, each class is associated with a price per transmitted bit, byte, or packet. Typically, transmitting a packet in a higher priority service class results in a higher cost but a better QoS (lower delay and loss probability).

In [152], an adaptive packet forwarding mechanism was proposed for a DiffServ network where video packets are mapped onto different DiffServ service levels. The authors in [153] proposed a rate-distortion optimized packet marking technique to deliver MPEG-2 video sequences in a DiffServ IP network. Their goal was to minimize the bandwidth consumption in the premium class while achieving nearly constant perceptual quality. In [12], cost-distortion optimized multimedia streaming over DiffServ networks was studied for preencoded media. However, the work in [12, 152, 153] does not incorporate video source coding.

A formulation as in (6.1) for transmitting video over a DiffServ network is studied in [3]. In this formulation, the network parameters c are the QoS class and the scheduling decision for each packet. The source parameters s are the quantization step size and prediction mode for each packet. The cost constraint C_0 comes from the negotiation between the user and the Internet service provider (ISP) through the service level agreement (SLA).

In this setting, different DS classes result in different QoS levels, such as different probabilities of packet loss and network delay. The random network delay for each packet is incorporated into the calculation of the probability of packet loss; this delay is managed through selecting the source-coding parameters and packet priority. More specifically, finer quantizers lead to higher video reconstruction quality but longer delay (given the same QoS class), which results in higher loss probability. In addition, the packet QoS class also needs to be selected in a way to properly balance cost, delay, and video quality. For example, packets that are hard to conceal but can be easily encoded should use coarser quantizers and a higher QoS class. Packets that are easily concealable can be sent using a lower QoS class. For packets that are hard to encode, the best choice may be a finer quantizer and a higher QoS. The goal is to find a *cost-distortion* optimized selection of parameters, which is achieved through solving (6.1). It results in the joint selection of source-coding parameters and QoS classes so that the received video quality and the overall cost are optimally balanced.

For the simulations in [3], four QoS classes are considered. The channel parameters for each QoS class are chosen based on a model-based TCP-friendly congestion controller as in [11, 48]. The costs for each QoS class are set proportional to the average throughput of the class, which takes into account the transmission rate, probability of packet loss, and network delay distribution. The details of how the parameters are chosen for each QoS class can be found in [3].

To illustrate the advantage of joint selection of the source-coding parameters and QoS class, we consider a reference system similar to today's Internet where only one QoS class is available. In this reference system, source-coding decisions are made to minimize the expected end-to-end distortion subject to the transmission delay constraint. The corresponding DiffServ system matches the delay and cost of the reference system. Simulations based on the parameter settings given in [3] showed that the proposed joint coding and packet classification approach outperforms the corresponding single-class reference system.

Figure 6.8 shows the temporal behavior of these two systems for one channel error realization using the Foreman test sequence at 30 fps. Figure 6.8(a) shows the PSNR per frame for one channel error realization of the minimum distortion approach, i.e., (6.1), and its corresponding reference systems. Figure 6.8(b) shows the cost per frame for the minimum cost approach, (6.2), and its corresponding reference systems. When the objective is to minimize the end-to-end distortion, the proposed DiffServ approach achieves an average PSNR improvement of 1.3 dB over the single-class reference system. Similarly, when considering the minimum cost formulation, an average cost savings of 20% per frame is achieved using the proposed technique.

In Fig. 6.9, the number of packets that the DiffServ system allocates to each service class is shown for each reference system. As shown in this figure, the performance gain is mainly due to the flexibility in choosing the QoS class per packet.

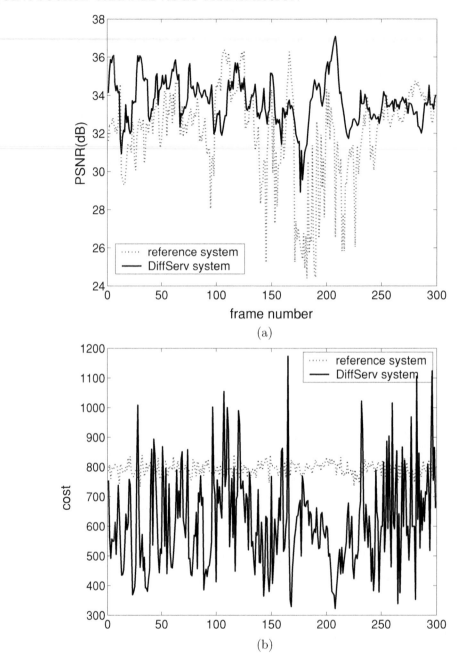

FIGURE 6.8: Comparison of the DiffServ approach with the single-class reference system: (a) minimum distortion approach, (b) minimum cost approach.

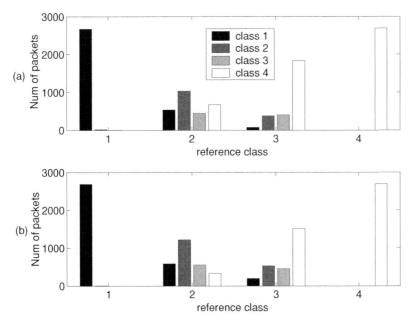

FIGURE 6.9: Distribution of packet classification in the DiffServ system: (a) minimum distortion approach, (b) minimum cost approach (adapted from [3]).

6.5 CONCLUSIONS

This chapter addressed the problem of joint source-channel coding for real-time video transmission over the Internet. We jointly considered error-resilient source coding at the encoder, various channel coding techniques at the application and/or transport layer, and error concealment at the receiver to achieve the best video quality. In particular, for the regular Internet, we have studied joint source coding and FEC and joint source coding and hybrid FEC/retransmission. We have also studied joint source coding and packet classification for video transmission over DiffServ networks. The objective was to provide examples to show that the general optimization framework presented in (6.1) can be applied to different video communications applications with different end-to-end delay requirement and network infrastructures.

CHAPTER 7

Wireless Video Transmission

Wireless video communications are a broad, active, and well-studied field of research [9, 10]. Different from Internet video transmission systems, wireless video transmission systems need adaptations at more layers, such as MAC retransmissions at the MAC or link layer, and modulation mode selection and power adaptation at the physical layer. In addition, transmission energy and computational energy play important roles in wireless video applications. In this chapter, we study the adaptation techniques for wireless video transmission using the general resource-distortion optimization framework presented in Chapter 3. We first provide the general problem formulation, and then study different adaptation techniques in different applications and network infrastructures, among which we pay more attention to the problem of joint source-channel coding and power adaptation.

7.1 INTRODUCTION

In an increasing number of applications, video is transmitted to and from portable wireless devices, such as cellular phones, laptop computers connected to wireless local area networks (WLANs), and cameras in surveillance and environmental tracking systems. For example, the dramatic increase in bandwidth brought by new technologies such as the present third generation (3G) and the emerging fourth generation (4G) wireless systems, and the IEEE 802.11 WLAN standards is beginning to enable video streaming capability in personal communications. Although wireless video communications are highly desirable in many applications, there are two major limitations in any wireless system, namely the hostile radio environment, including noise, time-varying channels and abundant electromagnetic interference, and dependence of mobile devices typically on a battery with a limited energy supply. Such limitations are especially of a concern for video transmissions because of the high bit rate and high energy consumption rate in encoding and transmitting video bitstreams. Thus, efficient use of bandwidth and energy becomes highly important in the deployment of wireless video applications.

To design an energy-efficient communications system, the first issue is to understand how energy is consumed in mobile devices. Generally speaking, energy in mobile devices is mainly used for computation, transmission, display, and driving speakers. Among those, computation and transmission are the two largest energy consumers. During computation, energy is used to run the operating system software and to encode and decode the audio and video signals. During transmission, energy is used to transmit and receive the RF audio and video signals. It should be acknowledged that computation has always been a critical concern in wireless communications. For example, energy-aware operating systems have been studied to efficiently manage energy consumption by adapting the system behavior and workload based on the available energy, job priority, and constraints. Computational energy consumption is especially a concern for video transmission, because motion estimation and compensation, forward and inverse DCT transformations, quantization, and other components in a video encoder all require a significant number of calculations. Energy consumption in computation has been recently addressed in [154], where a power-rate-distortion model has been proposed to study the optimal tradeoff between computation power, transmission rate, and video distortion. Nonetheless, advances in VLSI design and integrated circuit (IC) manufacturing technologies have led to ICs with higher and higher integration densities while using less and less power. According to Moore's law, the number of transistors on an IC doubles every 1.5 years. As a consequence, the energy consumed in computation is expected to become a less significant fraction of the total energy consumption. Therefore, in this chapter, we concentrate primarily on transmission energy.

In this chapter, we focus on the last hop of a wireless network. Specifically, we focus on the situation where video is captured and transmitted from a mobile wireless device to the base station, which is likely to be the bottleneck of the whole video transmission system. Such mobile wireless devices typically rely on a battery with a limited energy supply. In this case, the efficient utilization of transmission energy is a critical design consideration [155]. Thus, the problem is how to encode a video source and send it to the base station in an energy efficient way.

In addition, as discussed in Chapter 3, in order to deal with the time-varying channel, source coding and channel coding need to be jointly adapted to the changing channel conditions. Recently, several adaptation techniques have been proposed specifically for energy efficient wireless video communications [9,32,95]. A trend in this field of research is the joint adaptation of source coding and transmission parameters based on the time-varying source content and channel conditions, which generally requires a "cross-layer" optimization perspective [27]. Specifically, the lower layers in a protocol stack, which directly control transmitter power, need to obtain knowledge of the importance level of each video packet from the video encoder located at the application layer. In addition, it can also be beneficial if the source encoder is aware of the estimated CSI that is available at the lower layers, as well as which channel parameters at the lower layers can be controlled, so that it can make smart decisions in selecting the source-coding parameters to achieve the best video delivery quality. For this reason, joint consideration of video encoding and power control is a natural way to achieve the highest efficiency in transmission energy consumption.

Recall that the general goal of the resource-distortion optimization framework discussed in Chapter 3 requires the minimization of the end-to-end distortion while using a limited amount of resources and meeting the delay constraints as shown in (3.9), which is repeated below

$$\min_{s \in \mathcal{S}^M, c \in \mathcal{C}^M} E[D(s, c)]$$
$$\text{s.t.} \quad C(s, c) \leq C_0 \tag{7.1}$$
$$T(s, c) \leq T_0.$$

For energy-efficient wireless video transmission applications, we denote by C_0 the constraint on the total energy consumed in delivering the video sequence to the end user. Correspondingly, the channel-coding parameters would include more general channel adaptation parameters such as the transmission rate, physical-layer modulation modes, and the transmitter power.

In addition to the energy constraint, each video packet should meet a delay constraint in order to reach the receiver in time for playback. In order to efficiently utilize resources such as energy and bandwidth, those two adaptation components of transmission power and delay should be jointly designed. In this chapter, we discuss research efforts that focus on balancing energy efficiency with the above considerations.[1] The studies are carried out based on the general resource-distortion optimization framework in (7.1), as each adaptation technique for video transmission over wireless networks is a special case of it. Next we review the details.

7.2 JOINT SOURCE CODING AND FEC

As with Internet video transmission, the problem of joint source coding and FEC for wireless video communications focuses on the optimal bit allocation between source and channel coding by solving (3.8), which is a special case of (7.1) when energy is not involved. Different from Internet video transmissions, channel errors in wireless channels are typically in the form of bit errors, thus FEC is achieved by adding redundant bits within packets to provide intrapacket protection. RCPC and RS codes are widely used for this purpose. Recent studies have considered using turbo codes [146, 147] due to their ability to achieve capacity close to Shannon's bound [145]. As mentioned above, this topic has been extensively studied. Some examples are provided next.

---

[1] For simplicity, we will not provide details of the link-layer retransmission including MAC layer retransmission in this monograph. Instead, we assume that the function of link-layer retransmissions is disabled to avoid introducing extra latency. We further assume that physical-layer or link-layer FEC parameters can be accessed by the application layer so that they can be jointly decided with those in application layer and transport layer.

Optimal bit allocation has been studied in [98] based on a subband video codec. A binary symmetric channel (BSC) with an AWGN model has been considered for simulations. The source-coding parameters are the bit rate of the source subband and the channel-coding parameters are the FEC parameter for each subband. A similar problem has been studied for video transmission over a Rayleigh fading wireless channel in [63] based on an H.263+ SNR scalable video codec. In that work, universal RD characteristics (URDC) of the source scheme are employed to make the optimization tractable. Both studies use RCPC codes to achieve the intrapacket FEC. RS codes are used to perform channel coding in [85] for video transmission over a random BSC. Based on their proposed RD model, the source-coding parameter is the intra-MB refreshment rate and the channel-coding parameter is the channel rate.

An adaptive cross-layer protection scheme is presented in [27] for robust scalable video transmission over 802.11 wireless LANs. In this study, the video data are preencoded using an MPEG-4 FGS coder and thus source coding adaptation is not considered. The channel parameters considered in this study include adaptation components in various layers in the protocol stack, including the application-layer FEC (RS codes are used to achieve inter-packet protection), the MAC retransmission limit, and the packet sizes. These channel parameters are adaptively and jointly optimized for efficient scalable video transmission.

As for image transmission, joint source-channel coding algorithms have been studied in [156] for efficient progressive image transmission over a BSC. In this study, two unequal error protection schemes have been studied: a progressive rate-optimal scheme and a progressive distortion-optimal scheme. With a given transmission bit budget, in the progressive rate-optimal scheme, the goal is to maximize the average of the expected number of correctly received source bits over a set of intermediate rates. In the progressive optimal-distortion problem, the goal is to minimize the average of the expected distortion over a set of intermediate rates. Embedded coder such as the SPIHT coder [157] and JPEG2000 is used for source coding. As for channel coding, an RCPC code and a rate-compatible punctured turbo code are used for the rate-optimal scheme and distortion-optimal scheme, respectively. Thus, the source-coding parameter is the number of embedded bits allocated to each packet and the channel-coding parameter is the selected channel rates for the protection of each packet. Although the proposed schemes have slightly worse performance at a few rates close to the target rate as compared to the schemes that are maximizing the correctly received source bits or minimizing the expected distortion at the target transmission rate, the proposed schemes have better performance at most of the intermediate rates.

7.3   JOINT SOURCE CODING AND POWER ADAPTATION

Joint source coding and power allocation techniques deal with the varying error sensitivity of video packets by adapting the transmission power per packet based on the source content and

CSI. In other words, these techniques use transmission power as part of a UEP mechanism. In this case, the channel coding parameter is the power level for each video packet. Video transmission over CDMA networks using a scalable source coder (3D SPIHT), along with error control and power allocation is considered in [158]. A joint source coding and power control approach is presented in [159] for optimally allocating source coding rate and bit energy normalized with respect to the multiple-access interference noise density in the context of 3G CDMA networks. In [4], optimal mode and quantizer selection are considered jointly with transmission power allocation.

To illustrate some of the advantages of joint adaptation of the source coding and transmission parameters in wireless video transmission systems, we present experimental results which are discussed in detail in [4]. We compare a joint source coding and transmission power allocation (JSCPA) approach, i.e., the approach described by (7.1), with an independent source coding and power allocation (ISCPA) approach in which s and c are independently adapted. In Fig. 7.1, we plot the expected PSNR per frame of both approaches for the Foreman test sequence coded at 15 fps. It is important to note that both approaches use the same transmission energy and delay per frame.

As shown in Fig. 7.1, the JSCPA approach achieves significantly higher quality (expected PSNR) per frame than the ISCPA approach. Because the video encoder and the transmitter operate independently in the ISCPA approach, the relative importance of each packet, i.e., its contribution to the total distortion, is unknown to the transmitter. Therefore, the transmitter treats each packet equally and adapts the power in order to maintain a constant probability of packet loss. The JSCPA approach, on the other hand, is able to adapt the power per packet, and thus the probability of loss, based on the relative importance of each packet. For example, more power can be allocated to packets that are difficult to conceal. By correlating the actual video with Fig. 7.1, it can be seen that the PSNR improvement is the greatest during periods of high activity. For example, around frame 100 there is a scene change in which the camera pans from the Foreman to the construction site. During this time, the JSCPA approach achieves PSNR improvements of up to 3.5 dB. This gain comes from the JSCPA approach's ability to increase the power while decrease the number of bits sent in order to improve the reliability of the transmission. The ISCPA scheme is unable to adapt the protection level and thus incurs large distortion during periods of high source activity.

We show the visual quality comparison of the two approaches in Fig. 7.2. An expected reconstructed frame is shown from the Foreman sequence when the same amount of energy is consumed in the two approaches. It can be clearly seen that the JSCPA approach achieves a much better video reconstruction quality than the ISCPA approach. Next we plot in Fig. 7.3 the resource (transmission power and bits) allocation map for each MB in a frame in order to show where the gains come from by jointly considering source coding and power adaptation.

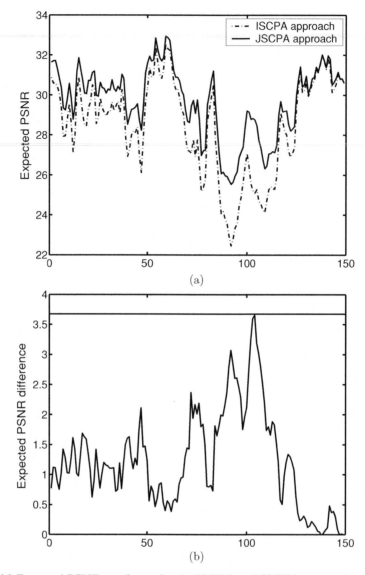

FIGURE 7.1: (a) Expected PSNR per frame for the ISCPA and JSCPA approaches (b) Difference in expected PSNR between the two approaches (adapted from [4]).

Figures 7.3(a) and 7.3(b) show frames 42 and 43 of the Foreman sequence, respectively. For frame 43, the two approaches can achieve the same expected video quality, but the JSCPA approach needs nearly 60% less energy to transmit this frame than the ISCPA approach. Figures 7.3(c) and 7.3(d) show the probability of loss for each packet in frame 43 for the JSCPA and ISCPA approaches, respectively. Darker MBs correspond to a smaller probability

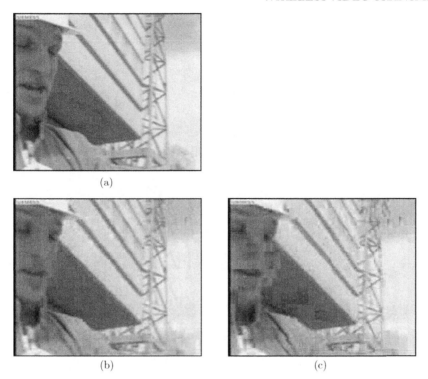

(a)

(b)   (c)

FIGURE 7.2: Frame 92 in the Foreman sequence. (a) original frame (b) expected frame at the decoder using the JSCPA approach and (c) expected frame at the decoder using the ISCPA approach (adapted from [4]).

of packet loss and MBs that are not transmitted are marked by white. As seen in Fig. 7.3(c), more protection is given to the region of the frame that corresponds to foreman's head. Therefore, more power is used to transmit this region as opposed to the background. As shown in Fig. 7.3(d), however, the ISCPA approach has fixed probability of loss, which means that the power used to transmit the region corresponding to the foreman's head is the same as the power used to transmit the background. Therefore, the ISCPA approach wastes energy by transmitting MBs in the background with the same power as macroblocks in the high activity region.

As for the source coding, in the ISCPA approach, the video encoder may allocate more bits to packets in high activity regions, as shown in Fig. 7.3(f). Because the transmission power is fixed in this approach, more energy is used to transmit packets with more bits, as shown in Fig. 7.3(h). Therefore, in the ISCPA approach, more energy may be allocated to high activity regions, but the likelihood of these regions being correctly received is the same as the background. In the JSCPA approach, the bit and power allocations are done jointly. Thus,

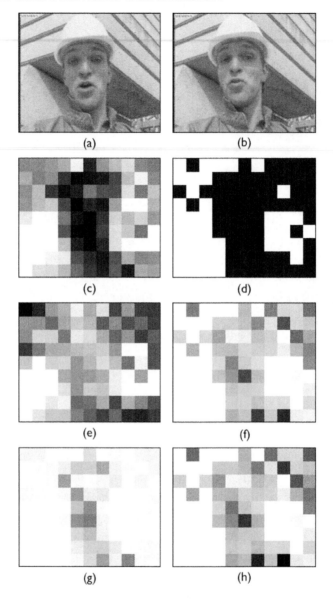

FIGURE 7.3: (a) Frame 42 and (b) Frame 43 in the original Foreman sequence. Probability of packet loss per macroblock for frame 43 using the (c) JSCPA approach, (d) ISCPA approach. Darker MBs correspond to a lower probability of packet loss. MBs that are not transmitted are shown in white. Bits per MB using the (e) JSCPA approach, (f) ISCPA approach. Darker MBs correspond to more bits. Transmission energy per MB using the (g) JSCPA approach, (h) ISCPA approach. Darker MBs correspond to higher transmission energy (adapted from [4]).

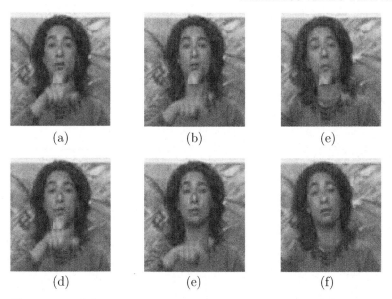

FIGURE 7.4: Illustration of the error propagation effect using the QCIF Silent test sequence. MD approach: frame number (a) 109, (b) 110, (c) 123. VAPOR approach: frame number (d) 109, (e) 110, (f) 123 (adopted from [5]).

the JSCPA approach is able to adapt the power per packet, making the probability of loss dependent on the relative importance of each packet, as shown in Figs. 7.3(e) and 7.3(g).

As mentioned in Chapter 2, the VAPOR (variance-aware per-pixel optimal resource allocation) approach is used to limit error propagation by accounting for not only the mean but also the variance of the end-to-end distortion [5]. In Fig. 7.4, we compare a series of reconstructed frames at the decoder for the minimum distortion (MD) approach (7.1) and the VAPOR approach using the same amount of transmission energy for the Silent sequence. These images are for a single-channel loss realization when the same MBs are lost in both schemes. We can clearly see the advantage of using the VAPOR approach. As shown in Fig. 7.4, both approaches suffer a loss in frame 109 where the woman's hand goes across her chin. The difference between the two approaches is that in frame 110, the VAPOR approach Intra refreshes this region while the MD approach does not. Thus, this error persists until frame 123 in the MD approach while it has been quickly removed by the VAPOR approach.

7.4   JOINT SOURCE–CHANNEL CODING AND POWER ADAPTATION

In this section, we study the resource-distortion optimization framework in the form of joint source-channel coding and power adaptation (JSCCPA), where error-resilient source coding, channel coding, and transmission power allocation are jointly designed to compensate for channel errors. This problem has been studied in [6, 37, 160]. In this case, the general channel-coding parameters control both channel coding and power allocations.

In [160], the study is based on progressive image and video transmission, and error-resilient source coding is achieved through optimized transport prioritization for layered video. Joint source-channel coding and processing power control for transmitting layered video over a 3G wireless network is studied in [37]. A joint FEC and transmission power allocation scheme for layered video transmission over a multiple user CDMA network is proposed in [158] based on the 3D-SPIHT codec. Source coding and error concealment are not considered in that work.

Source-channel coding and power adaptation can also be used in a hybrid wireless/wireline network, which consists of both wireless and wired links. An initial investigation of this topic is described in [6], where lower layer adaptation includes interpacket FEC at the transport layer and intrapacket FEC at the link layer, which are used to combat packet losses in the wired line and bit errors in the wireless link, respectively. In addition to channel coding, power is assumed to be adjustable in a discrete set at the physical layer. The selection of channel codes and power adaptation is jointly considered with source-coding parameter selection to achieve energy-efficient communication by solving (7.1). We next present this piece of work in detail. Note that we will also discuss video transmission over a single wireless channel, which is a special case of hybrid wireless networks where the wired link has no error and delay.

7.4.1   Product Code FEC

As discussed in Chapter 5.2, Reed–Solomon (RS) codes have good erasure correcting properties and thus have been widely used for providing inter packet protection. RCPC codes have been widely used for providing intrapacket protection, due to their flexibility in changing the code rate and simple implementation of both encoder and decoder. Thus, in this work, a systematic Reed–Solomon code is used to perform interpacket protection at the transport layer and RCPC codes to perform intrapacket protection at the link layer. The combination of the above two techniques, i.e., intra and interpacket FEC, is referred to as the product code FEC (PFEC).

In the PFEC scheme considered here, the first step is to perform RS coding at the transport layer. As with the packetization scheme described in Section 6.2, after packetization at this stage, each source packet is protected by an RS(N, M) code, as shown in Fig. 6.1. At the link layer, each packet (including the parity packets) is padded with parity bits. By using a particular RCPC code with rate r_k, the length of packet k is then $B_k = B_{s,k} + B_{c,k} = B_{s,k}/r_k$, where $B_{s,k}$ represents the source bits for the kth packet and $B_{c,k}$ is the link-layer channel bits for the same packet. Based on the described packetization scheme, we next discuss how to calculate the probability of packet loss.

Calculation of Transport Packet Loss Probability

We first discuss how to calculate the probability of loss for each transport packet. Let $\mathcal{Q}$, Γ, $\mathcal{R}$, and $\mathcal{P}$ be the sets of allowable source-coding parameters, RS coding parameter, RCPC coding parameters, and transmission power levels, respectively. Let $\mu_k \in \mathcal{Q}$, $\gamma \in \Gamma$, $\nu_k \in \mathcal{R}$, and $\eta_k \in \mathcal{P}$ represent the corresponding parameters selected for the kth packet. Let p_b be the BER after link-layer channel decoding, i.e., p_b is the BER as seen by the application. Assuming independent bit errors (i.e., the additive noise and fading are each i.i.d. and independent of each other), the loss probability for a transport packet in the wireless channel can be calculated as

$$\beta_k(\mu_k, \nu_k, \eta_k) = 1 - (1 - p_b)^{B_{s,k}}. \tag{7.2}$$

Let α denote the packet loss rate in the wired part. At the IP level, as in [161], the network can be modeled as the combination of two independent packet erasure channels: the wired part with a loss rate α and the wireless part with a loss rate β. Thus, the overall loss rate of a transport packet is then equal to

$$\epsilon_k(\mu_k, \nu_k, \eta_k) = \alpha + (1 - \alpha)\beta_k(\mu_k, \nu_k, \eta_k). \tag{7.3}$$

Calculation of Source Packet Loss Probability

As in (2.3), the expected distortion for each source packet is given by

$$E[D_k] = (1 - \rho_k)E[D_{R,k}] + \rho_k E[D_{L,k}], \tag{7.4}$$

where $E[D_{R,k}]$ and $E[D_{L,k}]$ are the expected distortion when the kth source packet is either received correctly or lost, respectively. Thus, in order to calculate the expected distortion for each source packet, we need to know its probability of loss, ρ_k. Next, we briefly discuss how to calculate ρ_k. Readers can refer to [6] for the details.

Let $\boldsymbol{\mu} = \{\mu_1, \mu_2, \ldots, \mu_M\}$ denote the vector of source-coding parameters for the M source packets, and $\boldsymbol{\nu} = \{\nu_1, \nu_2, \ldots, \nu_N\}$ and $\boldsymbol{\eta} = \{\eta_1, \eta_2, \ldots, \eta_N\}$ the vectors of RCPC coding rates and power levels for the N transport packets in a frame, respectively. As given in [6], let $Q_j^t(N)$, $j = 1, \ldots, N$, $t = 1, \ldots, \binom{N}{j}$ denote the tth subset with j elements of $Q(N) = \{1, \ldots, N\}$, and $\overline{Q_j^t}$ its complement. For example, if $N = 3$, then $Q(3) = \{1, 2, 3\}$, $Q_1^1(3) = \{1\}$, $Q_1^2(3) = \{2\}$, $Q_1^3(3) = \{3\}$, $Q_2^1(3) = \{1, 2\}$, $Q_2^2(3) = \{1, 3\}$, $Q_2^3(3) = \{2, 3\}$, $Q_3^1(3) = \{1, 2, 3\}$. Let $I_j(N, k) = \{Q_j^t \subseteq Q(N) | k \in Q_j^t(N), |Q_j^t| = j\}$, then the loss probability of a source packet can be written as

$$
\rho_k(\boldsymbol{\mu}, \gamma, \boldsymbol{v}, \boldsymbol{\eta}) = \sum_{j=N(\gamma)-M+1}^{N(\gamma)} P_{\mathrm{loss},k}(N(\gamma), j)
$$

$$
= \sum_{j=N-M+1}^{N} \sum_{Q_j^t \in I_j(N,k)} \left(\prod_{i \in Q_j^t} \epsilon_i \prod_{l \in \overline{Q_j^t}} (1 - \epsilon_l) \right), \tag{7.5}
$$

where $P_{\mathrm{loss},k}(N, j)$ is the probability that the kth packet is not correctly decoded by the RCPC decoder and the total number of transport packets that are not correctly received from the group of N packets is j.

Note that the calculation of ρ_k itself is rather complicated, since it not only depends on the source-coding parameter, intrapacket FEC parameter, and power level parameter selected for that packet, but also on the parameters chosen for all the other packets in the frame. This complicated interpacket dependency stems from two factors. The first complication is due to the fact that the loss probability of a transport packet $\epsilon_k(\mu_k, v_k, \eta_k)$ differs from packet to packet. The second complication is due to the interpacket dependency introduced by interpacket FEC. Together, these make the expected distortion for one packet depend on the parameters selected for all the packets in the same frame.

7.4.2   Problem Formulation

We first consider real-time video transmission from a mobile device to a receiver through a heterogeneous wireless network. Following the general resource-distortion optimization framework (7.1), by jointly considering error-resilient source coding $\boldsymbol{\mu}$, transport-layer FEC γ, link-layer FEC $\boldsymbol{v}$, power adaptation $\boldsymbol{\eta}$, and error concealment, the JSCCPA problem is formulated as

$$
\min_{\boldsymbol{\mu} \in \mathcal{Q}, \gamma \in \Gamma, \boldsymbol{v} \in \mathcal{R}, \boldsymbol{\eta} \in \mathcal{P}} \sum_{k=1}^{M} E[D_k(\boldsymbol{\mu}, \gamma, \boldsymbol{v}, \boldsymbol{\eta})]
$$

$$
\text{s.t.} \quad C = \sum_{k=1}^{N(\gamma)} B_k P_k(\eta_k)/R_T \leq C_0 \tag{7.6}
$$

$$
T = \sum_{k=1}^{N(\gamma)} B_k/R_T \leq T_0,
$$

where B_k and P_k are, respectively, the number of bits (including both source bits and channel bits) and the power level for the kth packet; M and N are, respectively, the number of source packets and the total number of transport packets in one frame; R_T is the transmission rate; and C_0 and T_0 are the energy and transmission delay constraints for the frame, respectively.

As a special case, we focus on the last hop of a wireless network and consider transmitting real-time video from a mobile device to the base station over a single wireless link; this is likely to be the bottleneck of the whole video transmission system. In Section 7.4.4, we show through simulations that interpacket FEC is not helpful in this case (at least for the cases we have simulated). Thus, in this special case, the JSCCPA problem for video transmission over a wireless link can be formulated as in (7.6) by disabling the transport-layer FEC. In this case, we have $\rho_k(\mu_k, \nu_k, \eta_k) = \beta_k(\mu_k, \nu_k, \eta_k)$, which means that the probability of loss for one packet depends only on the parameters selected for this packet. Recall that based on (2.4), when the error-concealment scheme discussed in Section 2.3.6 is used, the expected distortion for source packet k is given by

$$E[D_k] = (1 - \rho_k)E[D_{R,k}] + \rho_k(1 - \rho_{k-1})E[D_{C,k}] + \rho_k\rho_{k-1}E[D_{Z,k}],$$

where $E[D_{C,k}]$ and $E[D_{Z,k}]$ are the expected distortions after concealment when the previous packet is either received correctly or lost, respectively. Consequently, the expected distortion for one packet depends only on the parameters selected for this packet and its previous packet.

7.4.3 Solution Algorithm

In this section, we present solutions to (7.6) based on Lagrangian relaxation. Recall that as discussed in Section 3.5, the difficulty in solving the resulting minimization problem depends on the complexity of the interpacket dependencies. Due to the different complexities of the interpacket dependencies of (7.6) (and the special case when the transport-layer FEC is disabled), the resulting minimization problems can be efficiently solved using an iterative descent algorithm that is based on the method of alternating variables for multivariate minimization [103] and deterministic dynamic programming, respectively.

Lagrangian Relaxation
First, we formulate a Lagrangian dual for (7.6) by introducing Lagrange multipliers, $\lambda_1 \geq 0$ and $\lambda_2 \geq 0$, for the transmission energy and delay constraints, respectively. The resulting Lagrangian is

$$L(\boldsymbol{\mu}, \gamma, \boldsymbol{\nu}, \boldsymbol{\eta}, \lambda_1, \lambda_2) = E[D] + \lambda_1(C - C_0) + \lambda_2(T - T_0), \tag{7.7}$$

and the corresponding dual function is

$$g(\lambda_1, \lambda_2) = \min_{\mu \in \mathcal{Q}, \gamma \in \Gamma, \nu \in \mathcal{R}, \eta \in \mathcal{P}} L(\mu, \gamma, \nu, \eta, \lambda_1, \lambda_2). \tag{7.8}$$

Note that the Lagrangian may not be separable because the distortion for the kth packet, $E[D_k]$, may depend on the parameters chosen for the other packets. The dual problem to (7.6) is then given by

$$\max_{\lambda_1 \geq 0, \lambda_2 \geq 0} g(\lambda_1, \lambda_2). \tag{7.9}$$

Solving (7.9) will provide a solution to (7.6) within a convex hull approximation. Assuming that we can evaluate the dual function for a given choice of λ_1 and λ_2, a solution to (7.9) can be found by choosing the correct Lagrange multipliers. This can be accomplished by using a variety of methods such as cutting-plane or sub-gradient methods [89]. Alternatively, based on the observed structure of this problem, a heuristic approach is proposed in [6], which is considerably more efficient than the above-mentioned methods.

In the proposed solution, when one Lagrange multiplier is fixed, the dual problem becomes a one-dimensional problem, which can be easily solved by standard convex search techniques, such as the bisection method as illustrated in Chapter 3. Next, we consider evaluating the dual function in (7.8), given appropriate λ_1 and λ_2.

Minimization of Lagrangian

In (7.6), for given Lagrange multipliers, minimizing the resulting Lagrangian itself is still complicated due to the fact that the loss probability of one source packet depends on the operational parameters chosen for all the other packets. Hence, we solve the minimization problem by an iterative descent algorithm that is based on the method of alternating variables for multivariate minimization [103]. To be precise, for each RS code $\gamma \in \Gamma$ (i.e., we do an exhaustive search for γ), the RS block size is $N(\gamma)$, which is also the number of total transport packets in a frame. Then by adjusting one set of operational parameters for one packet at a time, while keeping constant those for the other packets until convergence, we can minimize the Lagrangian, $L(\mu, \gamma, \nu, \eta, \lambda_1, \lambda_2)$ in (7.7). In particular, let $x_k = \{\mu_k, \nu_k, \eta_k\}$ denote the vector of the source coding, intra-FEC channel coding, and power level selected for the kth packet, and $x = \{x_1, x_2, \ldots, x_N\}$ denote the parameters selected for the N packets.[2] Let $x^{(t)} = \{x_1^{(t)}, x_2^{(t)}, \ldots, x_N^{(t)}\}$, for $t = 0, 1, 2, \ldots$, be the parameter vector selected by optimization at step t, where $x^{(0)}$ corresponds to any initial parameter vector selected for the N packets.

---

[2]Note that for $k > M$, there is no associated source coding parameter μ_k defined, because these packets are parity packets. However, the number of source bits in those parity packets is determined by the maximum of the source bits in the source packets.

This can be done in a round-robin style, e.g., let $t_n = (t \mod N)$. If $i \neq t_n$, let $x_i^{(n)} = x_i^{(n-1)}$. Otherwise, for $i = t_n$, the following optimization is carried out:

$$x_i^{(t)} = \arg \min_{x^{(t)}} L\left(x_1^{(t)}, \ldots, x_{i-1}^{(t)}, x_i, x_{i+1}^{(t)}, \ldots, x_N^{(t)}, \lambda_1, \lambda_2\right). \tag{7.10}$$

The optimal operational parameter vector $\boldsymbol{x}^{(t)}$ is updated until the Lagrangian $L(\boldsymbol{x}^{(t)}, \gamma, \lambda_1, \lambda_2)$ converges. Convergence is guaranteed because the Lagrangian is nonincreasing and bounded below [48, 158]. The computational complexity mainly comes from the calculation of $\rho_k(\boldsymbol{\mu}, \gamma, \boldsymbol{v}, \boldsymbol{\eta})$ as in (7.5), which depends on the block size of the RS code, $N(\gamma)$. Note that the above iterative algorithm generates a set of optimal parameters of $(\boldsymbol{\mu}, \boldsymbol{v}, \boldsymbol{\eta})$ for a particular γ. The final optimal parameters $(\boldsymbol{\mu}, \gamma, \boldsymbol{v}, \boldsymbol{\eta})$ correspond to the minimum Lagrangian with one particular γ and its corresponding optimal parameters of $(\boldsymbol{\mu}, \boldsymbol{v}, \boldsymbol{\eta})$.

If the special case, where the transport-layer FEC is not included in (7.6), is considered, we can accurately (since the global optimal solution is guaranteed) and efficiently minimize the resulting Lagrangian by using DP due to the limited interpacket dependencies.[3] For simplicity, let C_k and T_k denote the transmission energy and transmission delay for packet k, respectively. The resulting Lagrangian can be expressed as $L(\boldsymbol{\mu}, \boldsymbol{v}, \boldsymbol{\eta}, \lambda_1, \lambda_2) = \sum_{k=1}^{M} J(k)$, where

$$J(k) = E[D_k] + \lambda_1 C_k + \lambda_2 T_k.$$

From (7.2) and (7.4), the cost of each packet $J(k)$ is a function of μ_k, v_k, η_k and $E[D_{L,k}]$. As shown in (2.4), if we employ the error-concealment strategy described in Chapter 2.3.6, the cost of the kth packet can be described as

$$J(k) = J(\mu_{k-1}, v_{k-1}, \eta_{k-1}, \mu_k, v_k, \eta_k).$$

The dual can then be evaluated via dynamic programming. The time complexity[4] of this scheme is $O(M \cdot |\mathcal{Q} \times \mathcal{R} \times \mathcal{P}|^2)$ [17].

7.4.4 Experimental Results

In the simulations, we choose an H.263+ codec to perform source coding, and consider the QCIF Foreman test sequence at a frame rate of 30 fps. In all experiments, we assume that the network delay is long enough to preclude retransmissions.

We use an RCPC code with generator polynomials (133, 171), mother code rate 1/2, and puncturing rate $G = 4$. This mother rate is punctured to achieve the 4/7, 2/3, and 4/5 rate

---

[3]Note that due to the use of interpacket FEC in (7.6), where the expected distortion for one packet depends on the parameters selected for all the packets in the same frame, DP is not applicable in (7.6).

[4]Note that here we only discuss the time complexity for evaluating the dual given particular Lagrange multipliers. The total time complexity depends on the number of iterations needed to find the correct Lagrange multipliers.

TABLE 7.1: Performance (BERs) of RCPC over a Rayleigh Fading Channel with Interleaving (cr Denotes the Channel Rate)

SNR (dB)	2	6	10	14	18	22
cr = 1/2	1.4×10^{-3}	2.2×10^{-5}	2.1×10^{-6}	2.4×10^{-7}	6.4×10^{-8}	2.8×10^{-9}
cr = 4/7	1.1×10^{-1}	5.3×10^{-4}	4.1×10^{-5}	1.1×10^{-5}	3.8×10^{-6}	1.3×10^{-6}
cr = 2/3	3.2×10^{-1}	7.4×10^{-3}	1.7×10^{-4}	3.5×10^{-5}	1.2×10^{-5}	4.2×10^{-6}
cr = 4/5	4.2×10^{-1}	4.0×10^{-2}	6.6×10^{-4}	1.1×10^{-4}	3.6×10^{-5}	1.2×10^{-5}

codes. At the receiver, soft Viterbi decoding is used in conjunction with BPSK demodulation. We present experiments on Rayleigh fading channels, and the channel parameter is defined as $\mathrm{SNR} = a\frac{E_b}{N_0}$, where a is the expected value of the square of the Rayleigh distributed channel gain. In the simulations, the bit error rates for the Rayleigh fading under the assumption of ideal interleaving were obtained experimentally using simulations, as shown in Table 7.1. The method for simulation can be found in [138, 139].

Video Transmission over Hybrid Wireless Networks

We first evaluate the performance of the proposed PFEC on a hybrid wireless network, which consists of both wired and wireless links. We fix the transmission power in this study. This is mainly due to the high computational complexity in calculating (7.6) if all operational components are included. In addition, this simplified case allows us to better analyze the potential of the proposed PFEC approach in providing UEP. For the transport-layer interpacket FEC, we choose $\Gamma = \{(9, 9), (11, 9), (13, 9), (16, 9)\}$ as the available RS coding set. Longer blocks not only complicate the computation of $\rho_k(\boldsymbol{\mu}, \gamma, \boldsymbol{v}, \boldsymbol{\eta})$, but also introduce longer delays. The transmission rate is set as $R_T = 360$ kbps in all simulations of this subsection.

In this experiment, we compare the performances of two UEP systems: (i) the UEP product code FEC (UEP-PFEC) and (ii) pure link-layer FEC (UEP-LFEC). The goal is to illustrate the advantage of PFEC. Both systems are UEP optimized using (7.6), where the PFEC system allows transport-layer RS coding but the LFEC system does not. The two systems have the same energy and transmission delay constraints.

We illustrate the performance of the two systems in Fig. 7.5, where we plot the average decoded PSNR for various average SNR values in the wireless link and packet loss rates α in the wired link. As shown in Fig. 7.5, with the above simulation setup, when α is small, LFEC is close to PFEC. However, as the wired link gets worse, PFEC starts to outperform LFEC by up to 2.5 dB. This improved performance is due to the use of cross-packet protection in

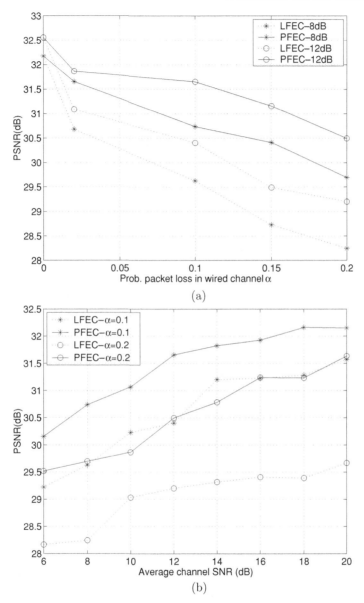

FIGURE 7.5: (a) PSNR versus α (b) PSNR versus average channel SNR, for PFEC and LFEC (adapted from [6]).

the transport layer. Table 7.2 shows how link-layer FEC rates are selected in the two systems. As can be seen, as the channel SNR improves, smaller link-layer protection is needed, i.e., higher channel rates are used. Second, compared to the LFEC approach, the PFEC approach uses lower rate codes of link-layer FEC because of the overhead from the transport-layer

TABLE 7.2: Link-Layer FEC Rates in Percentage in the UEP-PFEC and UEP-LFEC System (cr Denotes the Channel Rate)

α	SNR (dB)	LINK-LAYER FEC RATES IN PERCENTAGE IN UEP-PFEC				LINK-LAYER FEC RATES IN PERCENTAGE IN UEP-LFEC			
		CR=1/2	CR=4/7	CR=2/3	CR=4/5	CR=1/2	CR=4/7	CR=2/3	CR=4/5
	6	28.9	68.9	0.7	1.5	94.8	5.2	0	0
	8	26.7	5.2	68.1	0	63	8.9	28.1	0
0.1	10	3	0	89.6	7.4	28.9	0	71.1	0
	12	0.7	0	80.8	18.5	7.4	0	91.1	1.5
	16	0.7	0	23.7	75.6	0	0	80	20
	6	54.1	45.9	0	0	91.1	8.9	0	0
	8	17.8	10.4	71.8	0	47.4	11.9	40.7	0
0.2	10	2.2	0	97.8	0	23.7	0	76.3	0
	12	2	0	98	0	3	0	94	3
	16	1.5	0	11.8	86.7	0	0	60	40

FEC. In addition, it can be seen in the table that the link-layer FEC rates do not follow the change of α. This implies that the link-layer FEC is apparently less efficient than transport-layer FEC in reacting to packet losses in the wired link, because the link-layer FEC does not provide interpacket protection. Another observation from Fig. 7.5(a) is that when $\alpha = 0$, which corresponds to the case where the wired link is error free, the interpacket FEC in the transport layer becomes unnecessary and thus the optimized PFEC is equivalent to LFEC.

Additional simulations such as the comparison of UEP product FEC (adaptive link-layer FEC) and equal error protection (EEP) product FEC (fixed link-layer FEC) can also be found in [6].

Video Transmission over Wireless Links

In this study, we consider video transmission over a single wireless link. This can be regarded as a special case of hybrid wireless networks where the wired link is error and delay free. In the study above, we have shown that when the wired link has no errors ($\alpha = 0$), transport-layer interpacket FEC is not necessary; thus it is omitted, which makes the computations much more efficient. The goal of this subsection is to study the effectiveness of channel coding (intrapacket FEC) and power adaptation in achieving the optimal UEP. We focus on the tradeoff of the two adaptation components.

Performance Comparison of JSCPA and RERSC Systems

In this experiment, we compare the performance of two systems: the proposed framework in (7.6) with a fixed channel coding rate, which is referred to as JSCPA (joint source coding and power adaptation) system, and an RERSC (reference error resilient source coding) system which uses a fixed channel coding rate and transmission power. The transmission power levels can be adapted depending on the CSI and source content in the JSCPA system, but is fixed in the RERSC system. We refer to the RERSC system as the reference system, and evaluate it under different channel SNR (referred to as reference channel SNR) to generate the energy constraints for the JSCPA system. Thus, the two systems have the same transmission delay constraints and use the same amount of transmission energy.

We illustrate the performance of the two systems in Fig. 7.6(a), which shows the average decoded PSNR for the Foreman sequence under different channel SNR when $R_T = 360$ kbps, and Fig. 7.6(b), where we plot the average decoded PSNR versus transmission rate when the reference channel SNR is 12 dB. As shown in Fig. 7.6, by adjusting the power levels, the JSCPA system achieves a significant gain over the RERSC system. When the channel SNR is small, e.g., 8 dB, the gain can be as large as 6 dB in PSNR. The gain comes from the higher flexibility of the JSCPA approach, where the power level can be optimally assigned to different packets to achieve UEP for video packets that are of different importance. In addition, from Fig. 7.6, we can see that under some settings, JSCPA achieves little gain over RERSC [e.g., when the channel SNR is 12 dB and the channel coding rate used is low as shown in Fig. 7.6(b)]. This observation can help us assess the effective components in designing a practical video streaming system. Table 7.3 shows how transmission power is optimally selected for transmitting video packets in the proposed JSCPA system. The values inside the parentheses denote the percentage of packets with transmission power level 1, 2, 3, 4, 5, respectively. Note that the power level parameters 1, 2, 3, 4, and 5 are simplified substitutes for the actual transmission power values. The actual values are proportional to those parameters.

Performance Comparison of JSCCPA and JSCPA Systems

In the second experiment, we compare the performance of two systems: the proposed framework in (7.6) with a fixed link-layer channel coding rate, which is referred to as JSCPA (joint source coding and power adaptation) system, and the JSCCPA system as in (7.6). Note that the

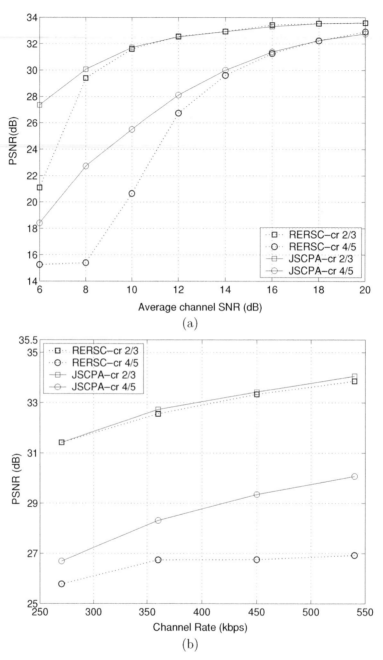

FIGURE 7.6: JSCPA versus RERSC (a) PSNR versus average channel SNR with $R_T = 360$ kbps (b) PSNR versus transmission rate with reference channel SNR be 12 dB (cr denotes channel rate in the legend) (adapted from [6]).

TABLE 7.3: Power Level Allocation of Power Level (1,2,3,4,5) in Percentage in the JSCPA System (the Reference Power Level is 3)

REFERENCE SNR (dB)	6	12	20
Channel rate = 1/2	(2.4,18.5,73.9,5.1,0)	(12.6,32.4,33.9,19.6,1.4)	(62.3,0,12.9,0,24.8)
Channel rate = 4/7	(18,0,14.3,66.1,1.6)	(2.3,29.9,56.4,11.0,0.3)	(10,35,39.2,13.4,2.3)
Channel rate = 2/3	(40,0,0,13,47)	(0.7,13.9,66.1,18.7.0.6)	(11.6,10.8,69.1,6.9,1.6)
Channel rate = 4/5	(45.8,0,0,0,54.2)	(2,4.1,41.8,47.3,4.9)	(8.2,31.5,43.8,15.3,1.3)

transport-layer FEC is disabled in both approaches. The transmission power levels can be adapted depending on the CSI and source content in both approaches. For the comparisons, the two systems have the same transmission delay constraints and use the same amount of transmission energy.

For the two systems, we plot the average decoded PSNR for the Foreman sequence under different channel SNRs (referred to as reference channel SNR) when $R_T = 360$ kbps in Fig. 7.7(a) and at different transmission rates when the reference channel SNR is 12 dB in Fig. 7.7(b). It can be seen that the JSCCPA approach achieves the upper bound of all JSCPA approaches. As expected, the gain comes from the higher flexibility of the JSCCPA approach, where channel-coding parameters can be optimally assigned to different packets to achieve UEP. Table 7.4 shows how channel coding rates are selected by the JSCCPA system. As can be seen, as the channel SNR improves, less channel protection is needed.

7.5   JOINT SOURCE CODING AND DATA RATE ADAPTATION

Joint source coding and transmission rate adaptation has also been studied as a means of providing energy efficient video communications. In order to maintain a certain probability of packet loss, the energy consumption increases as the transmission rate increases [162]. Therefore, in order to reduce energy consumption, it is advantageous to transmit at the lowest rate possible [163]. In addition to affecting energy consumption, the transmission rate determines the number of bits that can be transmitted within a given period of time. Thus, as the transmission rate decreases, the distortion due to source coding increases. Joint source coding and transmission rate adaptation techniques adapt the source-coding parameters and the transmission rate in order to balance energy consumption against end-to-end video quality.

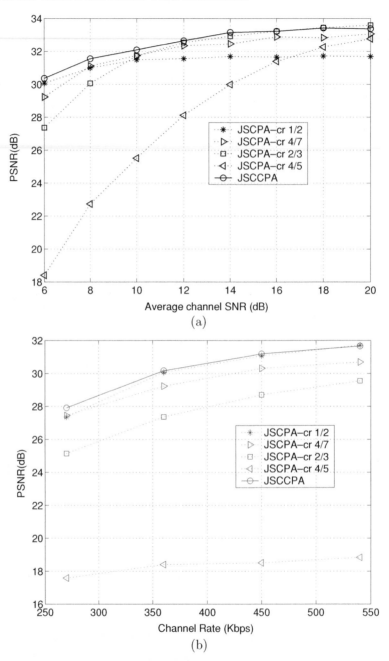

FIGURE 7.7: JSCCPA versus JSCPA (a) PSNR versus average channel SNR with $R_T = 360$ kbps (b) PSNR versus channel transmission rate with the reference channel SNR at 12 dB (cr denotes channel rate in the legend) (adapted from [6]).

TABLE 7.4: Channel Coding Rates in Percentage in JSCCPA System

REFERENCE SNR (dB)	6	8	10	12	14	16	18	20
Channel rate = 1/2	96.2	67.7	41.2	19.6	6.7	4.7	1.0	1.6
Channel rate = 4/7	3.8	31.9	57.3	69.6	61.3	35.0	17.8	5.6
Channel rate = 2/3	0	0.4	1.5	10.8	31.3	57.7	73.9	69.6
Channel rate = 4/5	0	0	0	0	0.7	2.6	7.3	23.2

In [101], the authors consider optimal source coding and transmission rate adaptation. In this work, each video packet can consist of a different number of MBs and can be transmitted at different rates. In addition, each packet has the option to stay idle at the transmitter instead of being transmitted immediately after it leaves the encoder buffer, when the current channel condition is very poor. The channel-coding parameters c then includes the selection of the number of MBs in each video packet, the transmission rate, and transmission schedule (waiting time at the transmitter) for each packet. The source coding parameters are the quantizer and prediction mode for each MB in each video packet. The goal is then to jointly select the source-coding and channel-coding parameters for each packet with the objective of minimizing the total expected energy required to transmit the video frame subject to both an expected distortion constraint and a delay per packet constraint, through solving the minimum cost optimization formulation presented in (6.2). In this work, stochastic dynamic programming is used to find an optimal source coding and transmission policy based on a Markov state channel model. A key idea in this work is that the performance can be improved by allowing the transmitter to suspend or slow down transmissions during periods of poor channel conditions, as long as the delay constraints are not violated.

7.6 CONCLUSIONS

In this chapter, we have studied JSCC for video transmission over wireless networks. Different from Internet video transmissions, the channel-coding parameters for wireless video transmission cover more general channel adaptation parameters such as the transmission rate, physical-layer modulation modes, and the transmitter power. In addition, efficient utilization of transmission energy is a critical design consideration for video transmission over wireless channels. In order to achieve the best video transmission quality with given resource constraints, various adaptation techniques should be jointly designed, including error-resilient source coding, transport-layer FEC, link-layer FEC, power adaptation, and error concealment.

We reviewed different adaptation strategies such as joint source coding and FEC, joint source coding and power adaptation, joint source-channel coding and power adaptation, and joint source coding and data rate adaptation. Among them we focused on two examples, joint source coding and power adaptation, and joint source-channel coding and power adaptation. The studies were carried out using the general resource-distortion optimization framework. Through the detailed discussion of the formulation construction, solution algorithms and simulation results, we explained the key technique concerns and illustrated the advantages of joint design of cross-layer adaptations. Although the current network protocols and infrastructure may not support all types of the above-studied cross-layer protocol communications, the optimal framework serves as a useful tool in the performance evaluation of suboptimal systems.

CHAPTER 8

Conclusions

8.1   DISCUSSION

According to Shannon's separation theory, source coding and channel coding can be separately designed and overall optimality can still be achieved. Source coding aims at removing redundancy from the source and achieving entropy, while channel coding aims at achieving error-free transmission by introducing redundancy. If the source rate (the entropy) is less than the channel capacity, then overall error-free transmission of the source is achievable; otherwise, the rate distortion theory gives the bounds of the theoretical lowest achievable distortion. Nonetheless, all the above hinges on ideal channel coding, i.e., the assumption of infinite complexity and delay, which, unfortunately, is not realistic for practical systems such as video communications.

Therefore, for a practical system, redundancy needs to be introduced during both source and channel encoding in a judicious way. In this monograph, we have addressed the theory and applications of JSCC for real-time video transmissions, and reviewed the advances in this area, i.e., the state-of-the-art implementations of JSCC in various network infrastructures and applications. We have used the term "channel encoding" in a general way to include modulation and demodulation, power adaptation, packet scheduling, and data rate adaptation.

Although technologies providing higher bit rates, lower error rates, and longer battery life of mobile devices are continuously being developed, higher QoS will almost certainly lead to higher user demands of service, which translates to higher resolution images of higher visual quality in video communications. In addition, although the most recent video coding standards H.263, MPEG-4, and H.264 provide a number of error resiliency tools, there are a number of resource allocation problems that need to be resolved in order to efficiently utilize those tools. It has not been our intention to cover all the realizations of JSCC for video transmission, but rather focus on illustrating techniques with specific emphasis on the joint adaptation of source and channel coding in the presented resource-distortion optimization framework. The overall topic presented here does not embody mature technologies yet. Instead, there is a plethora of issues that need to be addressed by considering new system structures.

8.2 FUTURE RESEARCH DIRECTIONS

Cross-layer design, currently a cutting edge approach in video transmission research, is a general term that encompasses JSCC. The basic concept embodied in cross-layer design is that the video transmission system needs to be adaptive to the changing network conditions to efficiently use limited network resources. In contrast, each layer is optimized or adapted to changing network conditions independently in the traditional layered protocol stack, where the adaptation is very limited due to the limited conversations between layers. Studies on cross-layer design have not only proved the necessity of joint design of multiple layers, but also identified future directions on network protocol suite developments to better support video communications over the current best effort networks.

Cross-layer design can be a powerful tool in that it can efficiently handle different types of channel errors in a heterogeneous IP-based wireless network that consists of both wired and wireless links. Cross-layer design of heterogeneous wireless Internet video systems is a relatively new and active field of research that warrants further investigation. Current work in this area includes the deployment of a proxy at the edge between the wired and wireless network to improve video quality [161]. Other studies have focused on end-to-end error control methods to deal with the unique characteristics of each network component [6, 164].

The resource-distortion framework presented in this monograph can be used to optimally adapt the source coding and network parameters given delay and cost constraints. Techniques embodying this framework all assume that these constraints for each frame (or group of frames) are set based on the application. Developing algorithms to set these constraints in a smart way is an important research direction. The underlying concept for those algorithms is similar to rate control, but distinguishable in that these schemes must effectively distribute both bandwidth (which concerns delay) and resources, such as energy, over time.

This monograph has primarily focused on efficient resource allocation for a single user in a unicast scenario. Efficient resource allocation for multiuser and multicast video transmission systems is an area that has recently been explored, but requires significant further study.

Multiuser data transmission has become an important area in the context of the developing 3G and post 3G wireless networks such as HSDPA [165] and IEEE 802.16 [166]. This is especially true for downlink data transmission (i.e., transmission from a wireless base station to mobile clients) to multiple users. Advanced wireless access technologies available in these networks allow for fast and frequent channel feedback, which in turn enable the use of multiuser diversity techniques that assign a greater portion of the available resources to users with better quality channels [29, 167]. The need to simultaneously meet a level of fairness across users has led to techniques such as proportionally fair scheduling [168], and more generally, gradient-based scheduling policies [169, 170]. Gradient-based scheduling policies define a user utility as a function of some QoS measure, and then maximize a weighted sum of the users' data rates where the weights are determined by the gradients of the utility function. In the case of delay-constrained video transmission, one approach has been to use the queue length at the transmission buffer, or the delay of the head-of-line (HOL) packet of the video stream as a QoS measure [171, 172]. This type of approach can be considered as source independent since it does not explicitly consider the source content when performing resource allocation across users. In [172], a time division multiplexing (TDM) type of scheme is proposed which uses a source dependent importance measure as well as the delay of the HOL packet to determine priority across users. An implementation of a source dependent gradient-based scheduling policy, which would work in TDM/CDMA and most TDM/OFDMA (orthogonal frequency division multiple access) type systems is proposed in [173].

An initial study of joint source coding and data adaptation scheme for downlink video transmission in a multiuser wireless network has been proposed in [174], aiming at achieving the best video transmission quality from the base station to multiple mobile users. Thus in addition to the joint design of source coding and data rate for a single user according to the changing channel conditions as in [101], a key idea in this work is that the system performance can be further improved through multiuser diversity, whereby transmissions of video packets for each user are optimally scheduled at each time slot according to the current channel condition, as long as the delay constraints are not violated. The study is carried out using a rate-distortion optimization formulation, where the source coding and data rate are jointly designed according to the changing channel conditions.

More work in this area is necessary to include packet loss considerations and with them, the possibility of combining packet scheduling with hybrid ARQ and FEC schemes. A natural extension of these methods is that of uplink video transmission (from multiple clients to a single-base station), in which case, issues such as centralized versus distributed resource allocation will also need to be addressed.

Although we have focused on single layer block-based motion compensated video coding, extending the JSCC framework to new coding paradigms, e.g., scalable video coding, distributed signal processing and sensor networks, is another promising research direction. In those regimes, the definition of cost and distortion must be tweaked appropriately based on the application. For example, to ensure network connectivity, energy constraints may be based on the remaining battery life of each node in the network.

Last but not least, one of the most critical aspects of designing efficient resource allocation algorithms is an accurate metric for end-to-end video quality evaluation. Among the relevant areas of research where significant work is necessary is the development of perceptually motivated objective quality metrics for packet-based video transmissions. By getting a better understanding of how the adaptation of certain source and network parameters affects the perceived video quality, algorithms can be designed to allocate limited resources more effectively.

Bibliography

[1] D. Wu, Y. T. Hou and Y.-Q. Zhang, "Transporting real-time video over the Internet: Challenges and approaches," *Proc. IEEE*, vol. 88, pp. 1855–1877, Dec. 2000.

[2] F. Zhai, Y. Eisenberg, T. N. Pappas, R. Berry and A. K. Katsaggelos, "Rate-distortion optimized hybrid error control for real-time packetized video transmission," *IEEE Trans. Image Process.*, vol. 15, pp. 40–53, Jan. 2006.

[3] F. Zhai, C. E. Luna, Y. Eisenberg, T. N. Pappas, R. Berry and A. K. Katsaggelos, "Joint source coding and packet classification for real-time video transmission over differentiated services networks," *IEEE Trans. Multimedia*, vol. 7, pp. 716–726, Aug. 2005.

[4] Y. Eisenberg, C. E. Luna, T. N. Pappas, R. Berry and A. K. Katsaggelos, "Joint source coding and transmission power management for energy efficient wireless video communications," *IEEE Trans. Circuits Sys. Video Technol.*, vol. 12, no. 6, pp. 411–424, June 2002.

[5] Y. Eisenberg, F. Zhai, T. N. Pappas, R. Berry and A. K. Katsaggelos, "VAPOR: Variance-aware per-pixel optimal resource allocation," *IEEE Trans. Image Process.*, vol. 15, pp. 289–299, Feb. 2006.

[6] F. Zhai, Y. Eisenberg, T. N. Pappas, R. Berry and A. K. Katsaggelos, "Joint source-channel coding and power adaptation for energy-efficient wireless video communications," *Signal Process.: Image Commun.*, vol. 20/4, pp. 371–387, April 2005.

[7] B. Haskell, "International standards activities in image data compression," in *Proc. Scientific Data Compression Workshop*, 1989, pp. 439–449, NASA Conf. Pub 302, NASA Office of Management, Scientific and Technical Information Division.

[8] S. Wenger, "H.264/AVC over IP," *IEEE Trans. Circ. Syst. Video Techn.*, vol. 13, pp. 645–656, July 2003, Special issue on the H.264/AVC video coding standard.

[9] B. Girod and N. Farber, "Wireless video," in *Compressed Video Over Networks*, A. Reibman and M.-T. Sun, Eds. New York: Marcel Dekker, 2000, pp. 124–133.

[10] D. Wu, Y. Hou and Y.-Q. Zhang, "Scalable video coding and transport over broad-band wireless networks," *Proc. IEEE*, vol. 89, pp. 6–20, Jan. 2001.

[11] Q. Zhang, W. Zhu and Y.-Q. Zhang, "Resource allocation for multimedia streaming over the Internet," *IEEE Trans Multimedia*, vol. 3, no. 3, pp. 339–355, Sept. 2001.

[12] A. Sehgal and P. A. Chou, "Cost-distortion optimized streaming media over Diffserv networks," in *Proc. IEEE Int. Conf. Multimedia and Expo*, Lausanne, Switzerland, Aug. 2002.

[13] Y. Wang, G. Wen, S. Wenger and A. K. Katsaggelos, "Review of error resilience techniques for video communications," *IEEE Signal Process. Mag.*, vol. 17, pp. 61–82, July 2000.

[14] K. Ramchandran and M. Vetterli, "Best wavelet packet bases in a rate-distortion sense," *IEEE Trans. Image Process.*, vol. 2, no. 2, pp. 160–175, Apr. 1993.

[15] K. Ramchandran, A. Ortega and M. Vetterli, "Bit allocation for dependent quantization with applications to multiresolution and MPEG video coders," *IEEE Trans. Image Process.*, vol. 3, pp. 533–545, Sept. 1994.

[16] A. Ortega and K. Ramchandran, "Rate-distortion methods for image and video compression," *IEEE Signal Process. Mag.*, vol. 15, pp. 23–50, Nov. 1998.

[17] G. M. Schuster and A. K. Katsaggelos, *Rate-Distortion Based Video Compression: Optimal Video Frame Compression and Object Boundary Encoding*, Kluwer, 1997.

[18] C. E. Shannon, "A mathematical theory of communication," *Bell Syst. Tech. J. 27*, pp. 379–423 and 623–656, July 1948.

[19] C. E. Shannon, "Coding theorems for a discrete source with a fidelity criterion," *IRE International Convention Records*, pp. 142–163, 1959, Part 4.

[20] K. Nichols, S. Blake, F. Baker and D. Black, "Definition of the differentiated services field (DS field) in the IPv4 and IPv6 headers," *RFC 2474, IETF*, Dec. 1998, http://www.rfc-editor.org/rfc/rfc2474.txt.

[21] B. E. Carpenter and K. Nichols, "Differentiated Services in the Internet," *Proc. IEEE*, vol. 90, no. 9, pp. 1479–1494, Sept. 2002.

[22] B. Girod, J. Chakareski, M. Kalman, Y. J. Liang, E. Setton and R. Zhang, "Advances in network-adaptive video streaming," in *Proc. Tyrrhenian Int. Workshop Digital Communications*, Capri, Italy, Sept. 2002, pp. 1–8.

[23] Q. Chen and K. P. Subbalakshmi, "Joint source-channel decoding for MPEG-4 video transmission over wireless channels," *IEEE J. Select. Areas Commun.*, vol. 21, pp. 1780–1789, Dec. 2003.

[24] P. A. Chou, A. E. Mohr, A. Wang and S. Mehrotra, "Error control for receiver-driven layered multicast of audio and video," *IEEE Trans. Multimedia*, pp. 108–122, March 2001.

[25] P. A. Chou and A. Sehgal, "Rate-distortion optimized receiver-driven streaming over best-effort networks," in *Proc. IEEE Int. PacketVideo Workshop*, Pittsburgh, PA, April 2002.

[26] H. Zheng, "Optimizing wireless multimedia transmissions through cross layer design," in *Proc. IEEE ICME*, Baltimore, MD, July 2003, vol. 1, pp. 185–188.

[27] M. var der Schaar, S. Krishnamachari, S. Choi and X. Xu, "Adaptive cross-layer protection strategies for robust scalable video transmission over 802.11 WLANs," *IEEE J. Select. Areas Commun.*, vol. 21, pp. 1752–1763, Dec. 2003.

[28] Q. Zhang, W. Zhu and Y.-Q. Zhang, "End-to-end QoS for video delivery over wireless Internet," *Proc. IEEE.*, vol. 93, pp. 123–134, Jan. 2005.

[29] S. Shakkottai, T. Rappaport and P. Karlsson, "Cross-layer design for wireless networks," *IEEE Commun. Mag.*, vol. 41, pp. 74–80, Oct. 2003.

[30] W. Kumwilaisak, Y. T. Hou, , Q. Zhang, W. Zhu, C.-C. J. Kuo and Y.-Q. Zhang, "A cross-layer quality-of-service mapping architecture for video delivery in wireless networks," *IEEE J. Select. Areas Commun.*, vol. 21, pp. 1685–1698, Dec. 2003.

[31] D. Wu and R. Negi, "Effective capacity: a wireless link model for support of qualify of service," *IEEE Trans. Wirel. Commun.*, vol. 2, pp. 630–643, July 2003.

[32] A. K. Katsaggelos, F. Zhai, Y. Eisenberg and R. Berry, "Energy efficient video coding and delivery," *IEEE Wirel. Commun.*, vol. 12, pp. 24–30, Aug. 2005.

[33] P. Pahalawatta and A. K. Katsaggelos, "Review of content aware resource allocation schemes for video streaming over wireless networks," *J Wirel. Commun. Mobile Comput.*, vol. 7, pp. 131–142, Feb. 2007.

[34] J. Postel, "RFC791: Internet protocol," RFC 791, Internet Engineering Task Force, Sept. 1981, http://www.rfc-editor.org/rfc/rfc791.txt.

[35] Douglas E. Comer, *Internetworking with TCP/IP*, vol. 1, Upper Saddle River, NJ: Prentice-Hall, 1995.

[36] L. Larzon, M. Degermark and S. Pink, "UDP Lite for real time multimedia applications," in *Proc. of the QoS miniconference of IEEE Int. Conf. Commun., (ICC'99)*, Vancouver, Canada, June 1999. <doi>Not Found</doi>

[37] Q. Zhang, W. Zhu and Y.-Q. Zhang, "Network-adaptive scalable video streaming over 3G wireless network," in *Proc. IEEE Int. Conf. Image Process. (ICIP)*, Thessaloniki, Greece, Oct. 2001.

[38] H. Schulszrinne, S. Casner, R. Frederick and V. Jacobson, "RTP: a transport protocol for real-time applications," RFC 3550, Internet Engineering Task Force, July 2003, http://www.networksorcery.com/enp/rfc/rfc3550.txt.

[39] D. D. Clark and D. L. Tennenhouse, "Architecture considerations for a new generation of protocols," *Comput. Commun. Rev.*, vol. 20, no. 4, pp. 200–208, Sept. 1990.

[40] M. Handley and C. Perkins, "Guidelines for writers of RTP payload format specifications," RFC 2736, IETF, Dec. 1999.

[41] D. Hoffman, G. Fernando and V. Goyal, "RTP payload format for MPEG1/MPEG2 video," *RFC 2038, Internet Engineering Task Force*, Oct. 1996, http://www.faqs.org/rfcs/rfc2038.html.

[42] Y. Kikuchi, T. Nomura, S. Fukunaga, Y. Matsui and H. Kimata, "RTP payload format for MPEG-4 audio/visual streams," *RFC 3016, Internet Engineering Task Force*, Nov. 2000.

[43] T. Turletti and C. Huitema, "RTP payload format for H.261 video streams," *RFC 2032, Internet Engineering Task Force*, Oct. 1996, http://www.faqs.org/rfcs/rfc2032.html.

[44] S. Wenger, M. M. Hannuksela, T. Stockhammer, M. Westerlund and D. Singer, "RFC3984 – RTP payload format for H.264 video," *RFC 3984, Internet Engineering Task Force*, Feb. 2005, http://www.faqs.org/rfcs/rfc3984.html.

[45] M. Gallant and F. Kossentini, "Rate-distortion optimized layered coding with unequal error protection for robust Internet video," *IEEE Trans. Circuits Syst. Video Technol.*, vol. 11, no. 3, pp. 357–372, March 2001.

[46] D. Wu, Y. T. Hou, W. Zhu, H.-J. Lee, T. Chiang, Y.-Q. Zhang and H. J. Chao, "On end-to-end architecture for transporting MPEG-4 video over the Internet," *IEEE Trans. Circuits Syst. Video Technol.*, vol. 10, pp. 923–941, Sept. 2000.

[47] S. Floyd, M. Handley, J. Padhye and J. Widmer, "Equation-based congestion control for unicast applications," *Tech Report*, International Computer Science Institutes, Berkley, CA, March 2000.

[48] P. A. Chou and Z. Miao, "Rate-distortion optimized streaming of packetized media," *IEEE Trans. Multimedia*, vol. 8, pp. 390–404, April 2006.

[49] R. Puri, K.-W. Lee, K Ramchandran and V. Bharghavan, "An integrated source transcoding and congestion control paradigm for video streaming in the Internet," *IEEE Trans. Multimedia*, vol. 3, pp. 18–32, March 2001.

[50] Y.-G. Kim, J.-W. Kim and C.-C. Jay Kuo, "TCP-friendly Internet video with smooth and fast rate adaptation and network-aware error control," *IEEE Trans. Circuit Syst. Video Technol*, vol. 14, pp. 256–268, Feb. 2004.

[51] Q. Zhang, W. Zhu and Y.-Q. Zhang, "Channel-adaptive resource allocation for scalable video transmission over 3G wireless network," *IEEE Trans. Circuit Syst. Video Technol.*, vol. 14, pp. 1049–1063, Aug. 2004.

[52] K. N. Ngan, C. W. Yap and K. T. Tan, *Video Coding for Wireless Communication Syst.*, New York: Marcel Dekker, 2001.

[53] H. Lin and S. K. Das, "Performance study of TCP/RLP/MAC in next generation CDMA systems," in *14th IEEE Int. Symp. Personal, Indoor, Mobile Radio Commun. (PIMRC)*, Beijing, China, Sept. 2003.

[54] F. Hartanto and H. R. Sirisena, "Hybrid error control mechanism for video transmission in the wireless IP networks," in *Proc. of IEEE 10th Workshop on Local and Metropolitan Area Networks (LANMAN'99)*, Sydney, Australia, Nov. 1999, pp. 126–132.

[55] D. N. Rowitch and L. B. Milstein, "On the performance of hybrid FEC/ARQ systems using rate-compatible punctured turbo (RCPT) codes," *IEEE Trans. Commun.*, vol. 48, pp. 948–959, June 2000.

[56] S. Falahati, A. Svensson, N. C. Ericsson and A. Ahlén, "Hybrid type-II ARQ/AMS and scheduling using channel prediction for downlink packet transmission on fading channels," in *Nordic Radio Symposium*, 2001.

[57] Y. Wang and Q.-F. Zhu, "Error control and concealment for video communications: a review," *Proc. IEEE*, vol. 86, no. 5, pp. 974–997, May 1998.

[58] G. Côté, S. Shirani and F. Kossentini, "Optimal mode selection and synchronization for robust video communications over error-prone networks," *IEEE J. Select. Areas Commun.*, vol. 18, pp. 952–965, June 2000.

[59] B. Girod and N. Färber, "Feedback-based error control for mobile video transmission," *Proc. IEEE*, vol. 87, pp. 1707–1723, Oct. 1999.

[60] R. O. Hinds, T. N. Pappas and J. S. Lim, "Joint block-based video source-channel coding for packet-switched networks," in *Proc. SPIE*, vol. 3309, pp. 124–133, Jan. 1998.

[61] R. Zhang, S. L. Regunathan and K. Rose, "Video coding with optimal inter/intra-mode switching for packet loss resilience," *IEEE J. Select. Areas Commun.*, vol. 18, pp. 966–976, June 2000.

[62] D. Wu, Y. T. Hou, B. Li, W. Zhu, Y.-Q. Zhang and H. J. Chao, "An end-to-end approach for optimal mode selection in Internet video communication: theory and application," *IEEE J. Select. Areas Commun.*, vol. 18, no. 6, pp. 977–995, June 2000.

[63] L. P. Kondi, F. Ishtiaq and A. K. Katsaggelos, "Joint source-channel coding for motion-compensated DCT-based SNR scalable video," *IEEE Trans. Image Process.*, vol. 11, pp. 1043–1052, Sept. 2002.

[64] F. Zhai, R. Berry, T. N. Pappas and A. K. Katsaggelos, "Rate-distortion optimized error control scheme for scalable video streaming over the Internet," in *Proc. IEEE Int. Conf. Multimedia Expo*, Bartimore, MD, July 2003.

[65] H. M. Radha, M. van der Schaar and Y. Chen, "The MPEG-4 fine-grained scalable video coding method for multimedia streaming over IP," *IEEE Trans. Multimedia*, vol. 3, pp. 53–68, March 2001.

[66] D. G. Sachs, R. Anand and K. Ramchandran, "Wireless image transmission using multiple-description based concatenated codes," in *Proc. SPIE Image Video Commun. Process.*, San Jose, CA, Jan. 2000, vol. 3974, pp. 300–311.

[67] J. Chakareski, E. Setton, Y. Liang and B. Girod, "Video streaming with diversity," in *Proc. IEEE Int. Conf. Multimedia Expo*, Baltimore, MD, July. 2003.

[68] B. J. Dempsey, J. Liebeherr and A. C. Weaver, "On retransmission-based error control for continuous media traffic in packet-switched networks," *Comput. Networks ISDN Syst.*, vol. 28, pp. 719–736, Mar. 1996.

[69] G. J. Wang, Q. Zhang, W. W. Zhu and Y.-Q. Zhang, "Channel-adaptive error control for scalable video over wireless channel," in *IEEE MoMuc 2000*, Oct. 2000.

[70] C. E. Luna, Video Quality and Network Efficiency Trade-Offs in Video Streaming Applications, *Ph.D. Thesis*, Northwestern University, Evanston, IL, June 2002.

[71] B. Braden, D. Clark and S. Shenker, "Integrated services in the Internet architecture," *RFC 1633, Internet Engineering Task Force*, June 1994, http://www.ifla.org/documents/rfcs/rfc1633.txt.

[72] L. Zhang, S. E. Deering, S. Shenker and D. Zappala, "RSVP: a new resource ReSerVation protocol," *IEEE Network*, vol. 7, pp. 8–18, Sept. 1993.

[73] S. Blake and et al., "An architecture for differentiated services," *RFC 2475, IETF*, Dec. 1998, http://www.rfc-editor.org/rfc/rfc2475.txt.

[74] A. Katsaggelos, F. Ishtiaq, L. P. Kondi, M.-C. Hong, M. Banham and J. Brailean, "Error resilience and concealment in video coding," in *Proc. Eur. Signal Process. Conf.(EUSIPCO)*, Rhodes, Greece, Sept. 1998, pp. 221–228.

[75] Y.-K. Wang, M. Hannuksela, V. Varsa, A. Hourunranta and M. Gabbouj, "The error concealment feature in the H.26L," in *Proc. IEEE Int. Conf. Image Process.*, Rochester, New York, Sept. 2002.

[76] H. R. Rabiee, H. Radha and R. L. Kashyap, "Error concealment of still image and video stream with multi-directional recursive nonlinear filters," in *Proc. IEEE Int. Conf. Acoustics, Speech Signal Process.*, Atlanta, GA, May 1996, pp. 37–40.

[77] ITU-T, *Video codec test model near-term*, H.263 *Test-Model Ad Hoc Group*, Oct. 1999, Version 11 (TMN11), Release 2.

[78] T. N. Pappas and R. J. Safranek, "Perceptual criteria for image quality evaluation," in *Handbook of Image and Video Process.*, A. C. Bovik, Ed. New York: Academic Press, 2000.

[79] J. Chen and T. N. Pappas, "Perceptual coders and perceptual metrics," in *Human Vision and Electronic Imaging VI*, B. E. Rogowitz and T. N. Pappas, Eds., San Jose, CA, Jan. 2001, vol. 4299 of *Proc. SPIE*, pp. 150–162.

[80] S. A. Karunasekera and N. G. Kingsbury, "A distortion measure for blocking artifacts in images based human visual sensitivity," *IEEE Trans. Image Process.*, vol. 4, pp. 713–724, 1995.

[81] Z. Wang, A. C. Bovik, H. R. Sheikh and E. P. Simoncelli, "Image quality assessment: from error visibility to structural similarity," *IEEE Trans. Image Process.*, vol. 13, no. 4, pp. 600–612, April 2004.

[82] H. R. Sheikh, M. F. Sabir and A. C. Bovik, "A statistical evaluation of recent full reference image quality assessment algorithms," *IEEE Trans. Image Process.*, vol. 15, no. 11, pp. 3440–3451, Nov. 2006.

[83] C. J. van der B. Lambrecht and O. Verscheure, "Perceptual quality measure using a spatio-temporal model of human visual system," in *Proc. SPIE*, San Jose, CA, 1996, vol. 2668, pp. 450–461.

[84] S. Winkler, A. Sharma and D. McNally, "Video quality and blockness metrics for multimedia streaming applications," in *Proc. Int. Symp. on Wireless Personal Multimedia Commun.*, Aalborg, Denmark, Sept. 2001, pp. 547–552.

[85] Z. He, J. Cai and C. W. Chen, "Joint source channel rate-distortion analysis for adaptive mode selection and rate control in wireless video coding," *IEEE Trans. Circuits Syst. Video Technol.*, vol. 12, pp. 511–523, June 2002.

[86] T. Wiegand, N. Färber and B. Girod, "Error-resilient video transmission using long-term memory motion-compensated prediction," *IEEE J. Select. Areas Commun.*, vol. 18, no. 6, pp. 1050–1062, June 2000.

[87] H. Yang and K. Rose, "Advances in recursive per-pixel estimation of end-to-end distortion for application in H.264," in *Proc. IEEE Int. Conf. Image Process. (ICIP)*, Genova, Sept. 2005.

[88] T. M. Cover and J. A. Thomas, *Elements of Information Theory*, New York: Wiley, 1991.

[89] D. Bertsekas, *Nonlinear Programming*, Belmont, MA: Athena Scientific, 1995.

[90] B. Sklar, *Digital Communications: Fundamentals and Applications*, Englewood Cliffs, NJ: Prentice-Hall, 2nd edition, 2001.

[91] N. Farvardin and V. Vaishampayan, "Optimal quantizer design for noisy channels: an approach to combined source-channel coding," *IEEE Trans. Inf. Theory*, vol. IT-38, pp. 827–838, 1987.

[92] F. Farvardin, "A study of vector quantization for noisy channels," *IEEE Trans. Inf. Theory.*, vol. 36, pp. 799–809, July 1990.

[93] A. Kurtenbach and P. Wintz, "Quantizing for noisy channels," *IEEE Trans. Commun. Technol.*, vol. COM-17, pp. 291–302, Apr. 1969.

[94] R. E. Van Dyck and D. J. Miller, "Transport of wireless video using separate, concatenated and joint source-channel coding," *Proc. IEEE*, vol. 87, pp. 1734–1750, Oct. 1999.

[95] A. K. Katsaggelos, Y. Eisenberg, F. Zhai, R. Berry and T. N. Pappas, "Advances in efficient resource allocation for packet-based real-time video transmission," *Proc. IEEE*, vol. 93, pp. 135–147, Jan. 2005.

[96] F. Zhai, Y. Eisenberg, T. N. Pappas, R. Berry and A. K. Katsaggelos, "Rate-distortion optimized product code forward error correction for video transmission over IP-based

wireless networks," in *Proc. Int. Conf. Acoustics, Speech and Signal Process.*, Montreal, Canada, June 2004.

[97]   G. Davis and J. Danskin, "Joint source and channel coding for Internet image transmission," in *Proc. SPIE Conf. Wavelet Applications of Digital Image Process. XIX*, Denver, CO, Aug. 1996.

[98]   G. Cheung and A. Zakhor, "Bit allocation for joint source/channel coding of scalable video," *IEEE Trans. Image Process.*, vol. 9, pp. 340–356, March 2000.

[99]   J. Kim, R. M. Mersereau and Y. Altunbasak, "Error-resilient image and video transmission over the Internet using unequal error protection," *IEEE Trans. Image Process.*, vol. 12, pp. 121–131, Feb. 2003.

[100]   Y. Pei and J. W. Modestino, "Multi-layered video transmission over wireless channels using an adaptive modulation and coding scheme," in *Proc. IEEE Int. Conf. Image Process.*, Thesaloniki, Greece, Oct. 2001.

[101]   C. E. Luna, Y. Eisenberg, R. Berry, T. N. Pappas and A. K. Katsaggelos, "Joint source coding and data rate adaption for energy efficient wireless video streaming," *IEEE J. Select. Areas Commun.*, vol. 21, pp. 1710–1720, Dec. 2003.

[102]   F. Zhai, Y. Eisenberg, T. N. Pappas, R. Berry and A. K. Katsaggelos, "Rate-distortion optimized hybrid error control for real-time packetized video transmission," in *Proc. IEEE Int. Conf. Commun. (ICC'04)*, Paris, France, June 2004.

[103]   R. Fletcher, *Practical Methods of Optimization*, New York: Wiley, 2nd edition, 1987.

[104]   D. P. Bertsekas, *Dynamic Programming: Deterministic and Stochastic methods*, Englewood Cliffs, NJ: Prentice-Hall, 1987.

[105]   T. Wiegand, G. J. Sullivan, G. Bjntegaard and A. Luthra, "Overview of the H.264/AVC video coding standard," *IEEE Trans. Circuits Syst. Video Technol.*, vol. 13, pp. 560–576, July 2003, Special issue on the H.264/AVC video coding standard.

[106]   T. Wiegand, H. Schwarz, A. Joch, F. Kossentini and G. J. Sullivan, "Rate-constrained coder control and comparison of video coding standards," *IEEE Trans. Circuit Syst. Video Technol.*, vol. 13, pp. 688–703, July 2003, Special issue on the H.264/AVC video coding standard.

[107]   J. M. Shapiro, "Embedded image coding using zerotrees of wavelet coefficients," *IEEE Trans. Signal Process.*, vol. 41, pp. 3445–3463, Dec. 1993.

[108]   A. Said and W. Pearlman, "A new, fast and efficient image codec based on set partitioning in hierarchical trees," *IEEE Trans. Circuit Syst. Video Technol.*, vol. 6, pp. 243–250, June 1996.

[109]   *JPEG-2000 VM3.1 A Software*, ISO/IECJTC1/SC29/WG1 N1142, Jan. 1999.

[110]   K. Shen and E. J. Delp, "Wavelet based rate scalable video compression," *IEEE Trans. Circ. and Syst. for Video Techn.*, vol. 9, pp. 109–122, Feb. 1999.

[111] Y.-Q. Zhang and S. Zafar, "Motion-compensated wavelet transform coding for color video compression," *IEEE Trans. on Circ. and Syst. for Video Techn.*, vol. 2, pp. 285–296, Sept. 1992.

[112] J. R. Ohm, "Three-dimensional subband coding with motion compensation," *IEEE Trans. Image Process.*, vol. 3, pp. 559–571, Sept. 1994.

[113] S. Choi and J. W. Woods, "Motion-compensated 3-D subband coding of video," *IEEE Trans. Image Process.*, vol. 8, pp. 155–167, Feb. 1999.

[114] M. Flierl and B. Girod, "Video coding with motion-compensated lifted wavelet transforms," *Signal Process.: Image Commun.*, vol. 19, pp. 561–575, August 2004.

[115] X. Li, "Scalable video compression via overcomplete motion compensated wavelet coding," *Signal Process.: Image Commun.*, vol. 19, pp. 637–651, August 2004.

[116] A. Secker and D. Taubman, "Lifting-based invertible motion adaptive transform (LIMAT) framework for highly scalable video compression," *IEEE Trans. Image Proc.*, vol. 12, pp. 1530–1542, Dec. 2003.

[117] J.-R. Ohm, "Motion-compensated wavelet lifting filters with flexible adaptation," in *Proc. Tyrrhenian International Workshop on Digital Commun.*, Capri, Italy, Sept. 2002.

[118] M. van der Schaar and H. Radha, "Unequal packet loss resilience for Fine-Granular-Scalability video," *IEEE Trans. Multimedia*, vol. 3, pp. 381–394, Dec. 2001.

[119] J. Goshi, R. E. Ladner, A. E. Mohr, E. A. Riskin and A. Lippman, "Unequal loss protection for H.263 compressed video," in *Proc. IEEE Data Compression Conference*, Snowbird, UT, March 2003, pp. 73–82.

[120] S. Wenger, G. Côté, M. Gallant and F. Kossentini, *H.263 Test Model Number 11, Revision 3*, Q-15-G-16rev3, 1999.

[121] A. Albanese, J. Blomer, J. Edmonds, M. Luby and M. Sudan, "Priority encoding transmission," *IEEE Trans. Inf. Theory*, vol. 42, pp. 1737–1744, Nov. 1996.

[122] ITU-T, Video coding for low bitrate communication, *ITU-T Recommendation H.263*, Jan. 1998, Version 2.

[123] *Coding of audio-visual objects, Part 2-visual: Amendment 4: streaming video profile*, ISO/IEC 14496-2/FPDAM4, July 2000.

[124] ITU-T, *Draft ITU-T recommendation and final draft international standard of joint video specification (ITU-T Rec. H.264/ISO/IEC 14 496-10 AVC*, JVT of ISO/IEC MPEG and ITU-T VCEG, JVT G050, 2003.

[125] Y. Wang, M. T. Orchard, V. Vaishampayan and A. R. Reibman, "Multiple description coding using pairwise correlating transforms," *IEEE Trans. Image Process.*, vol. 10, pp. 351–366, Mar. 2001.

[126] M. Karczewicz and R. Kurceren, "The SP- and SI-frames design for H.264/AVC," *IEEE Trans. Circ. Syst. Video Technol.*, vol. 13, pp. 637–644, July 2003, Special issue on the H.264/AVC video coding standard.

[127] T. Wiegand, M. Lightstone, D. Mukherjee, T. Campbell and S. K. Mitra, "Rate-distortion optimized mode selection for very low bit rate video coding and the emerging H.263 standard," *IEEE Trans. Circ. Syst. for Video Technol.*, vol. 6, pp. 182–190, April 1996.

[128] G. J. Sullivan and T. Wiegand, "Rate-distortion optimization for video compression," *IEEE Signal Processing Mag.*, vol. 15, pp. 74–90, Nov. 1998.

[129] R. O. Hinds, *Robust model selection for block-motion compensated video encoding*, Ph.D. Thesis, MIT, Cambridge, MA, June 1999.

[130] M. Budagavi and J. D. Gilson, "Multiframe video coding for improved performance over wireless channels," *IEEE Trans. Image Process.*, vol. 10, pp. 252–265, Feb. 2001.

[131] E. Maani, F. Zhai and A. K. Katsaggelos, "Optimal mode selection and channel coding for video transmission over wireless channels using H.264/AVC," in *Proc. IEEE Int. Conf. Acoustics, Speech and Signal Process.*, Honolulu, Hawaii, April 2007.

[132] D. A. Eckhardt, *An Internet-style approach to managing wireless link errors*, Ph.D. Thesis, Carnegie Mellon University, Pittsburgh, PA, May 2002.

[133] M. Yajnik, S. Moon, J. Jurose and et al., "Measurement and modeling of the temporal dependence in packet loss," *Tech. Rep.* 98-78, UMASS CMPSCI, 1998.

[134] V. Paxson and S. Floyd, "Wide area traffic: the failure of Poisson modeling," *IEEE Trans. Networking*, vol. 3, pp. 226–244, June 1995.

[135] G. Hooghiemstra and P. Van Mieghem, "Delay distributions on fixed Internet paths," *Tech. Rep. report 20011020*, Delft University of Technology, 2001.

[136] L. Ozarow, S. Shamai and A. D. Wyner, "Information theoretic considerations for cellular mobile radio," *IEEE Trans. Vehicular Technology*, pp. 359–378, May 1994.

[137] T. S. Rappaport, *Wireless communications principle and practice*, Prentice Hall, 1998.

[138] J. Hagenauer, "Rate-compatible punctured convolutional codes (RCPC codes) and their applications," *IEEE Trans. Commun.*, vol. 36, pp. 389–400, Apr. 1988.

[139] J. G. Proakis, *Digital Commun.*, New York: McGraw-Hill, Aug. 2000.

[140] E. N. Gilbert, "Capacity of a burst-noise channel," *Bell Syst. Tech. J.*, vol. 39, no. 9, pp. 1253–1265, Sept. 1960.

[141] H. Wang and N. Moayeri, "Finite state Markov channel – A useful model for radio comunication channels," *IEEE Trans. Veh. Technol.*, vol. 44, pp. 163–171, Feb. 1995.

[142] Q. Zhang and S. A. Kassam, "Finite-state Markov model for Rayleigh fading channels," *IEEE Trans. Commun.*, vol. 11, pp. 1688–1692, Nov. 1999.

[143] N. Celandroni and F. Potortì, "Maximizing single connection TCP goodput by trading bandwidth for BER," *Int. J. Commun. Syst.*, vol. 16, pp. 63–79, Feb. 2003.

[144] I. S. Reed and G. Solomon, "Polynomial codes over certain finite fields," *SIAM J. Appl. Math.*, vol. 8, pp. 300–304, 1960.

[145] C. Berrou, A. Glavieux and P. Thitimajshima, "Near Shannon limit error-correcting coding and decoding: Turbo codes," in *Proc. IEEE Int. Conf. Commun.*, Geneva, Switzerland, May 1993, pp. 1064–1070.

[146] B. Sklar and F. J. Harris, "The ABCs of linear block codes," *IEEE Signal Process. Mag.*, pp. 14–35, July 2004.

[147] C. Lee and J. Kim, "Robust wireless video transmission employing byte-aligned variable-len Turbo code," in *Proc. SPIE Conf. Visual Commun. Image Process.*, 2002.

[148] S. B. Wicker and V. K. Bhargava, *Reed-Solomon Codes and Their Applications*, New York: Wiley, Sept. 1999.

[149] X. Yang, C. Zhu, Z. Li, G. Feng, S. Wu and N. Ling, "Unequal error protection for motion compensated video streaming over the Internet," in *Proc. IEEE Int. Conf. Image Process.*, Rochester, New York, Sept. 2002.

[150] P. Luukkanen, Z. Rong and L. Ma, "Performance of 1XTREME system for mixed voice and data communications," in *Proc. IEEE Int. Conf. Commun.*, Helsinki, Finland, June 2001, pp. 1411–1415.

[151] A. Chockalingam and G. Bao, "Performance of TCP/RLP protocol stack on correlated fading DS-CDMA wireless links," *IEEE Trans. Veh. Technol.*, vol. 49, pp. 28–33, Jan. 2000.

[152] J. Shin, J. Kim and C.-C. Kuo, "Quality-of-service mapping mechanism for packet video in differentiated services network," *IEEE Trans. Multimedia*, vol. 3, no. 2, June 2001.

[153] D. Quaglia and J. C. De Martin, "Adaptive packet classification for constant perceptual quality of service delivery of video streams over time-varying networks," in *Proc. IEEE Int. Conf. Multimedia Expo*, Baltimore, MD, July 2003, vol. 3, pp. 369–372.

[154] Z. He, Y. Liang and I. Ahmad, "Power-rate-distortion analysis for wireless video communication under energy constraint," in *Proc. SPIE Visual Commun. Image Process.*, San Jose, CA, Jan. 2004.

[155] N. Bambos, "Toward power-sensitive network architectures in wireless communications: Concepts, issues and design aspects," *IEEE J. Select. Areas Commun.*, vol. 18, pp. 966–976, June 2000.

[156] V. Stanković, R. Hamzaoui, Y. Charfi and Z. Xiong, "Real-time unequal error protection algorithms for progressive image transmission," *IEEE J. Select. Areas Commun.*, vol. 31, pp. 1526–1535, Dec. 2003.

[157] A. Said and W. A. Pearlman, "A new, fast and efficient image codec based on set partitioning in hierarchal trees," *IEEE Trans. Circuits System Video Technol.*, vol. 6, pp. 243–250, June 1996.

[158] S. Zhao, Z. Xiong and X. Wang, "Joint error control and power allocation for video transmission over CDMA networks with multiuser detection," *IEEE Trans. Circ. Syst. for Video Technol.*, vol. 12, pp. 425–437, June 2002.

[159] Y. S. Chan and J. W. Modestino, "A joint source coding-power control approach for video transmission over CDMA networks," *IEEE J. Select. Areas Commun.*, vol. 21, pp. 1516–1525, Dec. 2003.

[160] S. Appadwedula, D. L. Jones, K. Ramchandran and L. Qian, "Joint source channel matching for wireless image transmission," in *IEEE Int. Conf. Image Process.*, Chicago, IL, Oct. 1998.

[161] G. Cheung, W.-T. Tan and T. Yoshimura, "Rate-distortion optimized application-level retransmission using streaming agent for video streaming over 3G wireless network," in *Proc. IEEE Int. Conf. Image Process.*, Rochester, New York, Sept. 2002.

[162] R. G. Gallager, "Energy limited channels: Coding, multi-access and spread spectrum," *Tech. Rep., M.I.T. LIDS-P-1714*, Nov. 1987.

[163] A. El Gamal, C. Nair, B. Prabhakar, E. Uysal-Biyikoglu and S. Zahedi, "Energy-efficient scheduling of packet transmissions over wireless networks," in *Proc. IEEE INFOCOM'02*, 2002.

[164] J. Chakareski and P. A. Chou, "Application layer error-correction coding for rate-distortion optimized streaming to wireless clients," *IEEE Trans. Commun.*, vol. 52, pp. 1675–1687, Oct. 2004.

[165] *High Speed Downlink Packet Access; Overall Description, 3GPP Std. TS 25.308 v7.0.0*, 2006.

[166] IEEE-SA Standards Board, *IEEE Standard for Local and Metropolitan Area Networks; Part 16: Air Interface for Fixed and Mobile Broadband Wireless Access Syst.*, 2006, IEEE 802.16e Standard.

[167] R. Knopp and P. Humblet, "Information capacity and power control in single-cell multiuser communications," in *Proc. IEEE Int. Conf. Commun.*, Seattle, WA, June 1995, vol. 1, pp. 331–335.

[168] A. Jalali, R. Padovani and R. Pankaj, "Data throughput of CDMA-HDR a high efficiency—high data rate personal communication wireless system," in *Proc. IEEE Vehicular Technol. Conf.*, Tokyo, Japan, May 2000.

[169] R. Agrawal, V. Subramanian and R. Berry, "Joint scheduling and resource allocation in CDMA systems," *IEEE Trans. Inf. Theory*, to appear.

[170] K. Kumaran and H. Viswanathan, "Joint power and bandwidth allocation in downlink transmission," *IEEE Trans. Wireless Commun.*, vol. 4, pp. 1008–1016, May 2005.

[171] P. Falconio and P. Dini, "Design and performance evaluation of packet scheduler algorithms for video traffic in the high speed downlink packet access," in *15th IEEE International Symposium on Personal, Indoor and Mobile Radio Communications*, Sept. 2004.

[172] G. Liebl, M. Kalman and B. Girod, "Deadline-aware scheduling for wireless video streaming," in *Proc. IEEE Int. Conf. Multimedia and Expo*, July 2005.

[173] P. Pahalawatta, R. Berry, T. Pappas and A. Katsaggelos, "Content-aware resource allocation and packet scheduling for video transmission over wireless networks," *IEEE J. Select. Areas Commun.*, vol. 25, pp. 749–759, May 2007.

[174] F. Zhai, Z. Li and A. K. Katsaggelos, "Joint source coding and data rate adaptation for multi-user wireless video transmission," in *Proc. IEEE Int. Conf. Multimedia and Expo*, Beijing, China, July 2007.

Author Biography

Fan Zhai joined Texas Instruments (TI), Dallas, TX, in 2004. He is currently with the Department of Digital Entertainment Products, DSP Systems, where he is responsible for algorithm development and key intellectual property identification and management in image/video post-processing. His primary research interests include image and video signal processing and compression, multimedia communications and networking, and multimedia analysis. He has authored more than thirty publications, including one book chapter and eight journals, in the area of video compression and communications. He also holds one issued patent with five additional disclosures pending in the area of video post-processing, all of which have been implemented in TI's key video processing products. He has been on technical program committees for numerous prestigious international conferences including IEEE International Conference on Communications (ICC), 2006, IEEE Globecom, 2006, IEEE Consumer Communications and Networking Conference (CCNC), 2007, IEEE International Conference on Computer Communications and Networks (ICCCN), 2007, and IEEE International Conference on Multimedia and Expo (ICME) 2007.

Dr. Zhai received the B.S. and M.S. degrees in electrical engineering from Nanjing University, Nanjing, Jiangsu, China, in 1996 and 1998, respectively, and the Ph.D. degree in electrical and computer engineering from Northwestern University, Evanston, IL, in 2004.

Aggelos K. Katsaggelos received the Diploma degree in electrical and mechanical engineering from the Aristotelian University of Thessaloniki, Greece, in 1979 and the M.S. and Ph.D. degrees both in electrical engineering from the Georgia Institute of Technology, in 1981 and 1985, respectively. In 1985 he joined the Department of Electrical Engineering and Computer Science at Northwestern University, where he is currently professor. He was the holder of the Ameritech Chair of Information Technology (1997–2003). He is also the Director of the Motorola Center for Seamless Communications and a member of the Academic Affiliate Staff, Department of Medicine, at Evanston Hospital.

Dr. Katsaggelos has served the IEEE in many capacities (i.e., current member of the Publication Board of the IEEE Proceedings, editor-in-chief of the IEEE Signal Processing Magazine 1997–2002, member of the Board of Governors of the IEEE Signal Processing Society 1999–2001, and member of the Steering Committees of the IEEE Transactions on Image Processing 1992–1997). He is the editor of Digital Image Restoration (Springer-Verlag

1991), co-author of Rate-Distortion Based Video Compression (Kluwer 1997), co-editor of Recovery Techniques for Image and Video Compression and Transmission, (Kluwer 1998), co-author of Super-resolution for Images and Video (Claypool, 2007) and Joint Source-Channel Video Transmission (Claypool, 2007). He is the co-inventor of twelve international patents, a Fellow of the IEEE, and the recipient of the IEEE Third Millennium Medal (2000), the IEEE Signal Processing Society Meritorious Service Award (2001), an IEEE Signal Processing Society Best Paper Award (2001), and an IEEE International Conference on Multimedia and Expo Paper Award (2006). He is a Distinguished Lecturer of the IEEE Signal Processing Society (2007–08).

Printed in Great Britain
by Amazon

84167929R00086